AIRBUS A320
MCDU Operations

This is for training and entertainment only. For real flight, please see the Airbus manuals.

Introduction

Welcome to the most complete manual about the MCDU operations based on the FMS system of the great A320. This manual describes all functions of the MCDU (Multi-Function Control and Display Unit) for Airbus A320 including definitions, normal operations and abnormal operations in real flights.

Learn all about each part of the MCDU, each key, each function and every detail you need as a pilot.

After learning the all theory concepts, you will learn to operate the MCDU in different flights, including domestic flights, international flight and abnormal flights with emergencies.

At the end of this book, you will be ready for operating the MCDU like a professional pilot.

Index

Chapter 1

MCDU General Information

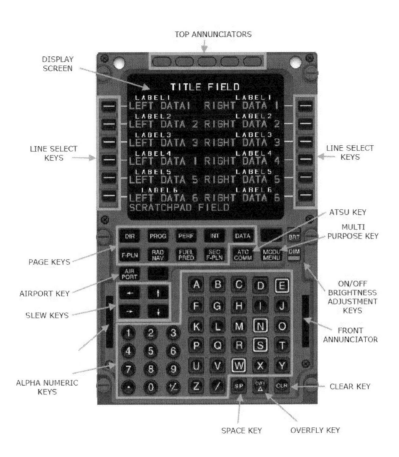

This is for training and entertainment only. For real flight, please see the Airbus manuals.

General information

The Flight Management Guidance System (FMGS) contains the following units:

○ Two Flight Management Guidance Computers (FMGC)

○ Two Multipurpose Control and Display Units (MCDU) (third MCDU optional)

○ One Flight Control Unit (FCU)

○ Two Flight Augmentation Computers (FAC).

The Flight Management and Guidance System (FMGS) provides predictions of flight time, mileage, speed, economy profiles and altitude. It reduces cockpit workload, improves efficiency, and eliminates many routine operations generally performed by the flight crew. The Flight Management Guidance System (FMGS) operates as follows:

○ During cockpit preparation the flight crew uses the Multipurpose Control and Display Unit (MCDU) to insert a preplanned route from origin to destination. This route includes SID, EN ROUTE, WAYPOINTS, STAR, APPROACH, MISSED APPR, and ALTN route as available from the navigation database.

○ Subsequently the system defines a vertical profile and a speed profile, taking into account ATC requirements and performance criteria.

Either FMGC performs all operations, if one FMGC fails.

The FMGS computes the aircraft position continually, using stored aircraft performance data and navigation data. Therefore it can steer the aircraft along a preplanned route and vertical and speed profiles. This type of guidance is said to be "managed".

If the flight crew wants to modify any flight parameter (SPD, V/S, HDG, etc.) temporarily, they may do so by using the various Flight Control Unit (FCU) selectors. The FMGS then guides the aircraft to the target value of this parameter that they have selected. This type of guidance is said to be "selected".

The two available types of guidance, then, are:

- Managed guidance guides the aircraft along the preplanned route and the vertical and speed/Mach profile. (The FMGS computes the target values of the various flight parameters).

- Selected guidance guides the aircraft to the target values of the various flight parameters the flight crew selects by using the FCU selectors.

Selected guidance always has priority over managed guidance.

Flight Management Guidance Computer (FMGC)

Each FMGC is divided into two main parts:

The Flight Management (FM) part controls the following functions:

- Navigation and management of navigation radios
- Management of flight planning
- Prediction and optimization of performance
- Display management.

The Flight Guidance (FG) part performs the following functions:

- Autopilot (AP) command
- Flight Director (FD) command
- Autothrust (A/THR) command.

10

Each FMGC has its own set of databases. The individual databases can be independently loaded into their respective FMGC, or independently copied from one FMGC to the other. Each FMGC contains these main databases:

The Navigation database (2.8 Mbytes) contains standard navigation data: Navaids, waypoints, airways, enroute information, holding patterns, airports, runways, procedures (SIDs, STARs, etc.), company routes, alternates.

The airline updates this part every 28 days, and is responsible for defining, acquiring, updating, loading, and using this data. The updating operation takes 20 min to complete or 5 min if cross loaded from the opposite FMGC.

The Airline Modifiable Information (AMI), also described as the FM Airline Configuration file, contains:

o Airline policy values: THR RED altitude, ACC altitude, EO ACC altitude, PERF factor, IDLE factor.

o Fuel policy values: Fuel for taxi, % of route reserve, maximum and minimum values of route reserve, etc.

o AOC functions customization.

The Aircraft Performance database includes the Engine model, Aerodynamical model, and Performance model. The airline cannot modify this database.

The Magnetic Variation database.

Each FMGC contains elements stored by the flight crew that enable them to create 20 waypoints, 10 runways, 20 navaids, and 5 routes.

Multipurpose Control and Display Unit (MCDU)

Two MCDUs are installed on the pedestal for flight crew loading and display of data. The use of the MCDU allows the flight crew to interface with the FMGC by selection of a flight plan for lateral and vertical trajectories and speed profiles. The flight crew may also modify selected navigation or performance data and specific functions of Flight Management (revised flight plan, engine-out, secondary flight plan, etc.). Additional data from peripherals (Centralized Fault Display System (CFDS), ARINC Communication Addressing and Reporting System (ACARS), Air Traffic Service Unit (ATSU)...) can also be displayed. Data that is entered into the MCDU that is illogical or beyond the aircraft capabilities will either be disregarded or will generate an advisory message.

System operation

The FMGS has three modes of operation:

- Dual mode (the normal mode)

- Independent mode. Each FMGC being controlled by its associated MCDU

- Single mode (using one FMGC only).

Dual Mode

This is the normal mode. The two FMGCs are synchronized: each performs its own computations and exchanges data with the other through a crosstalk bus. One FMGC is the master, the other the slave, so that some data in the slave FMGC comes from the master. All data inserted into any MCDU is transferred to both FMGCs and to all peripherals.

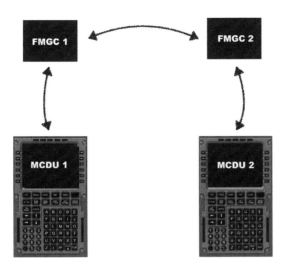

Master FMGC Logic

- If one autopilot (AP) is engaged, the related FMGC is master:

    - It uses the onside FD for guidance

    - It controls the A/THR

    - It controls the FMA 1 and 2.

- If two APs are engaged, FMGC1 is master.

- If no AP is engaged, and

    - The FD1 pb is on, then FMGC1 is master

    - The FD1 pb is off, and FD2 pb on then FMGC2 is master.

- If no AP/FD is engaged, A/THR is controlled by FMGC1.

Independent Mode

The system automatically selects this degraded mode under specific abnormal conditions (e.g. different database validity on both FMGCs). Both FMGCs work independently and are linked only to peripherals on their own sides of the flight deck ("onside" peripherals).

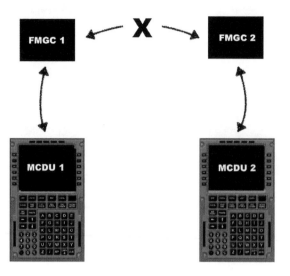

When this occurs, the "INDEPENDENT OPERATION" message is displayed on both MCDU scratchpads. Each MCDU transmits data it receives from its onside FMGC. It affects only the onside EFIS (Electronic Flight Instrument System) and RMP (Radio Management Panel).

On the POS MONITOR page (and GPS MONITOR page [?]), FMGS position (and GPS position [?]) from the opposite FMGC is not displayed. On the RAD NAV page (and PROG page, if the FMGS GPS is not installed), navaids tuned on the opposite MCDU are not displayed. Corresponding fields are blank.

Single Mode

The system automatically selects this degraded mode when one FMGC fails. When this occurs, the failed FMGC displays "OPP FMGC IN PROCESS" in white on the MCDU scratchpad. The corresponding ND displays the "SELECT OFFSIDE RNG/MODE" amber message.

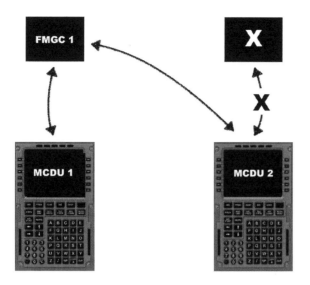

Both POS MONITOR pages display the same position (operative FMGC position). Both FDs are driven by the same FMGC. Any entry on either MCDU is sent to the operative FMGC.

General information of MCDU

The Multipurpose Control and Display Unit (MCDU) is a cathode ray tube that generates 14 lines of 24 characters each, including:

- A title line that gives the name of the current page in large letters

- Six label lines, each of which names the data displayed just below it (on the data field line)

- Six data field lines that display computed data or data inserted by the flight crew

- The scratchpad line that displays:

    - Specific messages

15

MCDU INTERFACE

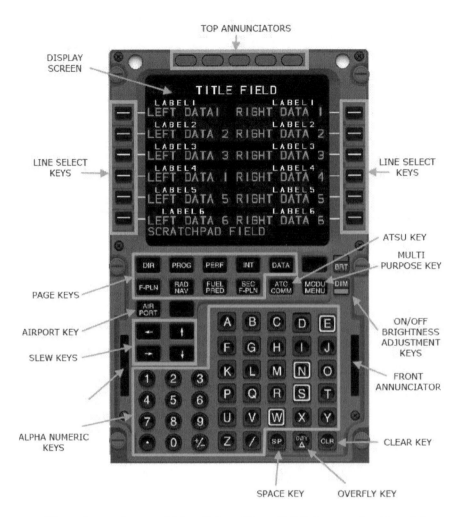

There is a column of Line Select Keys (LSKs) on each side of the screen. The flight crew uses these keys to:

○ Move a parameter they have entered in the scratchpad to the appropriate line on the main screen

○ Call up a specific function page indicated by a prompt displayed on the adjacent line

○   Call up lateral or vertical revision pages from the flight plan page.

KEYBOARD

The keyboard includes:

Function and Page Call up functions and pages the flight crew uses for flight management keys functions and computations.

↑ ↓ **(or SLEW)** keys Move a page up or down to display portions that are off the screen.

← → **keys** Moves to the next page of a multi-page element. An arrow in the top right corner indicates that another page is available.

AIRPORT key Calls up the flight plan page that contains the next airport along the current flight plan. Successive pushes on the key show the alternate airport, the origin airport (before takeoff), and the next airport again.

Number and letter keys allow the flight crew to insert data in the scratchpad so that they can use a line select key to enter it in the main display.

Three keys have special functions:

CLR (clear) key: Erases material (messages or inserted data) from the scratchpad or from certain areas of displayed pages

OVFY (overfly) key: Allows the aircraft to overfly a selected waypoint.

SP (space) key: Allows to insert a space in specific message.

Annunciators on the side of the keyboard

FAIL (amber): Indicates that the Multipurpose Control and Display Unit (MCDU) has failed.

MCDU MENU (white): Indicates that the flight crew should call up a peripheral linked to the MCDU (such as ACARS, ATSU or CFDS).

FM (white) Comes on while the flight crew is using the MCDU to display peripherals. This light tells the flight crew that the FMGC has an important message to deliver. The flight crew accesses the message by pressing the MCDU MENU key and the line select key adjacent to the FMGC prompt.

Annunciators on the top of the keyboard

FM 1 and FM 2 (amber): The onside FM is failed

IND (amber): The onside FM detects an independent mode of operation while both FM are healthly.

RDY (green): MCDU has passed its power up test after its BRT knob was turned to OFF.

BRT knob: Controls the light intensity of the entire MCDU.

Data Entry

The flight crew enters data by typing it into the scratchpad on the MCDU. Next, pressing the line select key (LSK) will load the data from the scratchpad into the desired field. An error message displays if the data is out of range or not formatted correctly. To correct data, the flight crew may clear the message with the clear (CLR) key and then retype the message into the scratchpad. Pressing the CLR key when the scratchpad is empty displays "CLR". To clear data from a field, select CLR from the scratchpad to the data field to be cleared.

18

MCDU Entry Format

The flight crew enters information into the MCDU at the bottom line of the scratchpad. When data has lead zeros, they may be omitted if desired. For example a three-digit wind direction of 060 may be typed as 60. The display will still show 060. To enter an altitude below 1 000 ft, the lead zero must be added as 0400 for 400 ft. This differentiates the altitude from a flight level.

To enter a double data entry such a speed/altitude, the separating slash must be used. If entering only the first part of a double entry, omit the slash. To enter only the second part of a double entry, a leading slash must be used i.e. /0400.

Messages

The scratchpad displays various messages for flight crew information. Theses messages are prioritized by importance to the flight crew as either amber or white.

Amber messages are:

* Navigation messages

* Data entry messages

* EFIS repeat messages.

Amber messages are categorized into two types:

* Type 1 message that is a direct result of a flight crew action. Type 1 messages are displayed immediately in the scratchpad ahead of other messages.

* Type 2 messages inform the flight crew of a given situation or request a specific action. Stored in "last in", "first out" message queue that holds maximum of 5 messages.

19

- Type 2 messages are displayed in the scratchpad only if there are no Type 1 messages or other data and will remain until all the messages have been viewed and cleared with the CLR key.

White messages are advisory only.

Characters

Small and large fonts are displayed according to the following rules:

- The title line and the scratchpad are displayed in large font

- Datafields are usually displayed in large font

- Label lines are displayed in small font

- Flight crew entries and modifiable data are displayed in large font

- Defaulted/computed and non modifiable data are displayed in small font.

Code of Colors

DATA	MCDU COLOR
TITLES, COMMENTS, <, >, ↑ ↓, ← →, DASHES, MINOR MESSAGES	WHITE
- MODIFIABLE DATA - SELECTABLE DATA - BRACKETS	BLUE
- NON MODIFIABLE DATA - ACTIVE DATA	GREEN
- MANDATORY DATA (BOXES) - FLIGHT CREW ACTION REQUIRED - IMPORTANT MESSAGES - MISSED CONSTRAINT	AMBER
- CONSTRAINTS - MAX ALTITUDE	MAGENTA
PRIMARY F-PLN	GREEN WAYPOINTS, WHITE LEGS
TEMPORARY F-PLN	YELLOW WAYPOINTS, WHITE LEGS
SECONDARY F-PLN	WHITE WAYPOINTS AND LEGS
MISSED APPROACH (not active)	BLUE WAYPOINTS, WHITE LEGS
ALTERNATE F-PLN (not active)	BLUE WAYPOINTS, WHITE LEGS
OFFSET	GREEN WAYPOINTS, WHITE LEGS, OFST DISPLAYED IN THE TITLE OF THE F-PLN PAGE
TUNED NAVAID	BLUE
"TO" WAYPOINT AND DESTINATION	WHITE

20

MCDU Pages

The information of the Multipurpose Control and Display Unit (MCDU) is divided in different pages for each phase of flight and each system.

MCDU Menu Page

This page lists the various systems which the pilot can access via the MCDU. The pilot selects a system by pressing the key adjacent to the name of that system. The name of the selected system is displayed in green, all others in white. If the MCDU cannot establish communication with the selected system, it displays "OUT". When a system calls for the pilot's attention, the MCDU displays "REQ" next to the system's name, and the "MCDU MENU" annunciator lights up. When the pilot presses the key next to the name of the system requiring attention, the "MCDU MENU" annunciator light goes out.

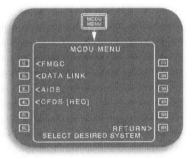

MCDU INIT Page

Calls up the flight plan initialization A page, which also gives the flight crew access to the B page. The flight crew uses the INIT pages to initialize Flight Management for the flight. The flight crew uses the INIT A page primarily to insert his flight plan and to align the inertial reference system.

21

The flight crew uses the INIT B page to insert aircraft weight, fuel on board, CG and various fuel requirements. The FMGS uses this data to compute predictions and fuel planning parameters. The flight crew has access to the INIT A page only in the preflight phase. INIT B page (not accessible after engine start) is called up by pressing the "NEXT PAGE" key.

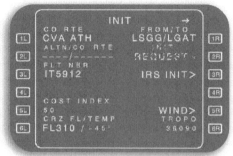

CO RTE: If the flight crew enters a company route number, the screen displays all data associated with that route (8 or 10 characters, depending on the pin program). Inserting the CO RTE into the RTE selection page also enters the CO RTE number in this field.

ALTN/CO RTE: This field is dashed, until a primary destination is entered in the 1R field. If a preferred alternate is associated with the primary destination, it is displayed in this field with the company route identification. The crew may manually enter an alternate and company route. If preferred alternate is not associated with the primary destination, NONE is displayed in this field.
When the alternate route and the primary destination do not match, the MCDU scratchpad displays "DEST/ALTN MISMATCH". If the primary destination is changed, this field is modified accordingly.

FLT NBR: The flight number automatically appears in this field, if it is stored with the company route. The flight crew may modify it, or enter a new number here.

COST INDEX: This is usually stored in the database along with the company route. The flight crew may modify it, or enter a new value here. It defaults to the last entered value, if a value is not stored in the database.

CRZ FL/TEMP: The cruise flight level is usually stored in the database along with the company route. If not, it has to be entered manually. If no cruise flight level is entered, the system will not furnish predictions, while the aircraft is on the ground.

The flight crew has to enter the temperature at cruise flight level in order to refine the predictions. Otherwise, these are computed for ISA conditions. (If no sign is entered, the system uses a plus)

FROM/TO: This field allows the pilot to enter a city pair (ICAO codes for city of origin and destination). This entry automatically deletes any previously entered company route and calls up the route selection page. If one airfield of the pair is not in the database, the display changes to the NEW RWY page.

IRS INIT: This field displays this legend only if the LAT and LONG fields are filled in, and at least one of the inertial reference systems is in ALIGN status (IRS in NAV position and alignment process not over). If the pilot presses this key when its field is displaying this legend, the present coordinates are sent to the IRSs and this completes the alignment process. If one of the three IRSs indicates an ALIGN FAULT occurrence, the prompt becomes REALIGN IRS.

WIND: The pilot presses this key in order to gain access to the climb wind page, unless a temporary flight plan exists. In this case, the scratchpad displays TEMPORARY F-PLN EXISTS.

TROPO: The default tropopause altitude is 36 090 ft. The pilot can use this field to modify it (60 000 ft maximum).

Route selection page

This page displays all the company routes, stored in the database, that are associated with the inserted city pair. They can be called up manually, or displayed automatically.

Manually: The pilot presses the FROM/TO or ALTN key on the INIT A page when a city pair is displayed.

Automatically: The system displays it, when the pilot enters a city pair, or defines an alternate on the INIT A page of the active or secondary flight plan.

TITLE: Idents for the city pair inserted on the INIT A page.
(The numbers in the upper righthand corner are the total number of company routes from this city pair stored in the database).

L1: This field shows the name of the company route. NONE appears, if there is no company route for this city pair.

RETURN: The pilot presses this key to return to the INIT A page.

INSERT: The pilot presses this key to insert the displayed company route in the flight plan, and return to the INIT A page.

Wind Page

Winds in climb, cruise, descent and approach are necessary to provide the pilot with reliable predictions and performance. Wind pages enable the pilot to enter and/or review the winds propagated by the FMGS or sent by ACARS for the various flight phases.

Climb Wind Page

This page enables the pilot to enter and/or review predicted wind vectors (direction and velocity) at up to 5 different levels.

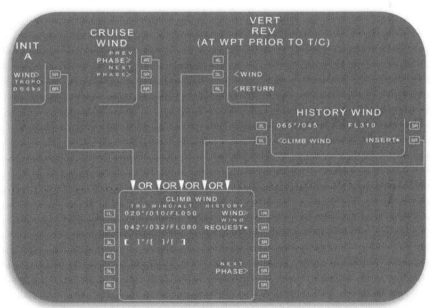

L1 to L5: This field displays the winds, entered at various climb altitudes. In blue before climb phase activation, and in green after climb phase activation.

This field may also display history winds or uplink winds. Large blue brackets are displayed before any wind entry. Pilot-entered and uplinked winds are displayed in large font. History wind data is displayed in small font. Upon sequencing the top of climb, the climb winds are deleted.

HISTORY WIND: Displayed in preflight phase only. This key calls up the history wind page. This page is not modifiable (small green font), but can be inserted into the CLIMB WIND page by using the 6R key and modified accordingly.

Cruise Wind Page

This page displays the wind direction and velocity for each cruise waypoint. The cruise wind page enables the definition of a temperature at a given altitude, and is accessed as follows:

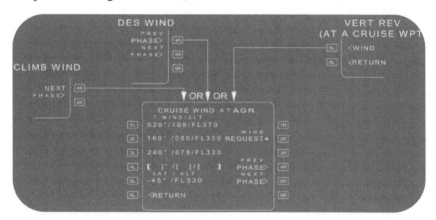

26

Descent Wind Page

This page enables the pilot to define and display the winds used for computing the descent profile. The pilot calls it up by selecting NEXT PHASE on the CRUISE WIND page, or the WIND prompt on the VERT REV page.

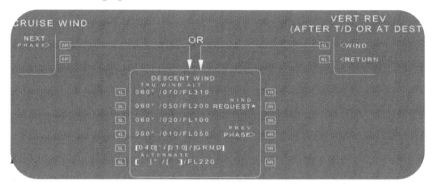

The pilot uses this page to initialize the gross weight and center of gravity, before starting the engines. The pilot can call it up from the INIT A page by pressing the NEXT PAGE key on the MCDU console, as long as engines have not been started. This page automatically reverts to the FUEL PRED page after the first engine is started.

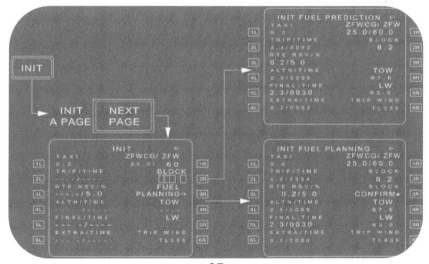

TAXI: This is the taxi fuel, which defaults to a preset value, (usually 200 kg or 400 lb). The pilot can change the value through this field.

TRIP/TIME: This field displays trip fuel and time when predictions become available. The pilot cannot modify this data.

RTE RSV%: This field displays the reserve fuel for the route and the corresponding percentage of trip fuel. It may be blank, if such is the policy of the operator. The pilot can either enter a route reserve, or a percentage, and the system then automatically computes the nominal value.

ALTN/TIME: This field displays alternate trip fuel and time, assuming that the Cost Index = 0 and that the aircraft flies at the default cruise flight level. This field displays its information in small font, and it cannot be modified by the flight crew.

TINAL/TIME: This field displays hold fuel and time, associated with continued flight to the alternate airport (or destination airport if no alternate is defined). The pilot may enter a final fuel or time (at alternate or destination) and the system will compute associated holding fuel/time available. Assumptions include a racetrack pattern 1 500 ft above the alternate airport, with the aircraft in CONF 1 at maximum endurance speed (or in accordance with the airline fuel policy established in the database).

EXTRA/TIME: This field displays the amount of extra fuel and the available time it represents for holding over the alternate or primary destination, if the pilot did not define an alternate.

EXTRA FUEL = BLOCK - (TAXI + TRIP + RSV + ALTN + FINAL)

The field displays this information in small font, and it cannot be modified by the flight crew.

ZFWCG/ZFW: The zero fuel weight and the location of the zero fuel weight CG are mandatory entries that allow the system to compute speed management and predictions. The pilot can modify this data.

BLOCK: The block fuel in this field is a mandatory entry that allows the system to predict the estimated fuel on board (EFOB).
When the pilot enters a block fuel, the page title changes to INIT FUEL PREDICTION. The FMGC may also compute the block, if the pilot selects the FUEL PLANNING function.

TOW: This field displays the computed takeoff weight. The pilot cannot modify it (small font).

LW: This field displays the computed landing weight at the primary destination. The pilot cannot modify it (small font).

TRIP WIND: This field allows the entry of a mean wind component for the trip from the primary origin to the primary destination. Upon entry of a CO RTE or FROM/TO pair, this field defaults to HD 000 in small blue font.

An entry preceded by H, HD is considered as headwind, +, T, TL as tailwind. The entered velocity is displayed in large blue font.
As soon as the crew inserts a wind on the CLIMB, CRUISE or DESCENT WIND page, the system no longer considers the trip wind.

MCDU FUEL PRED Page

The pilot presses the FUEL PRED key on the MCDU console to display fuel prediction information at destination and alternate, as well as fuel management data after the engines are started.

AT/UTC/EFOB: TIME is displayed before takeoff. UTC predictions are displayed after takeoff. After the pilot enters an Estimated Takeoff Time (ETT), the UTC is displayed.

GW/CG: GW: The system continually updates gross weight during the flight. If no zero fuel weight has been entered, the screen displays amber boxes next to this key. The pilot must enter information in these boxes in order to obtain a speed profile, speed computations, and predictions. The field displays dashes, as long as the system is not calculating fuel on board.

CG: The system continually updates the center of gravity location along the flight path. If no center of gravity has been entered, the screen displays amber boxes next to this key. The pilot must enter information in these boxes in order to obtain a speed profile, speed computations and predictions. The pilot can modify both the GW and the CG.

FOB: This field displays the fuel on board, which is computed: With information from the fuel flow and fuel quantity sensor (FF + FQ), or, from FF only (enter/FF to deselect FQ), or, from FQ only (enter/FQ to deselect FF). The pilot can modify this number.

MCDU F-PLN Page

These pages display all waypoints of the active and alternate flight plans, along with associated predictions. The pilot can make all revisions to the lateral and vertical flight plans from these pages:
He presses the left key to revise the lateral flight plan, and the right key to revise the vertical flight plan. He presses the F-PLN key on the MCDU console to access the page A of the active flight plan.

F-PLN Page A:

Page A displays time, speed, and altitude predictions for each waypoint of the active flight plan.

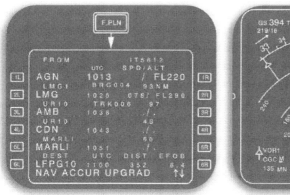 

FROM/UTC/SPD/ALT: These lines display consecutive waypoints along with associated predictions of time, speed or Mach and altitude for each. TIME is displayed before takeoff, and UTC after takeoff. After the pilot enters an estimated takeoff time (ETT), UTC is displayed. The time and flight level display at the FROM waypoint (first line of the flight plan) are values that the system memorized at waypoint sequencing.

SPD/ALT: The field dedicated to SPEED or MACH is blank at the FROM waypoint, except at the departure airport. (V1 associated with runway elevation, is displayed).

DEST UTC/TIME DIST/EFOB: DIST is the distance to destination along the displayed flight plan. EFOB is the estimated fuel on board at destination. The sixth line is permanent and is displayed in white font once predictions are available, except when a TMPY F-PLN is displayed or in some cases when an ALT CSTR is entered ("*CLB or DES*" prompt appears).

The generic flight plan page displays the FROM waypoint (last waypoint to be overflown) on the first line, and the TO waypoint (in white) on the second line. The FROM/TO flight plan leg is called the active leg. The flight crew can use the scroll keys to review all flight plan legs down to the last point of the alternate flight plan. The AIRPORT key serves as a fast slew key. The pilot can press it to call up the next airport (DEST, ALTN, ORIGIN) to be displayed on the flight plan page.

In order to return to the beginning of the flight plan page, the pilot presses the F-PLN key on the MCDU console. The display shows the name of the leg between two waypoints, and the distance between them on a line between the lines that identify them. During an approach, this in-between line also defines the angle of the final descent path. For example, "2-3 °" indicates that the leg is two nautical miles long, and the flight path angle is -3 °. The display shows the bearing between FROM and TO waypoints as the bearing from the aircraft position to the TO waypoint. It shows track (TRK) between the waypoints shown in lines 2 and 3. This is the outbound track of the next leg. If the database contains a published missed approach procedure, or if someone has inserted one manually, the display shows it in blue after the destination runway identification. It turns green when the go-around phase becomes active. After the last waypoint of the missed approach, the display shows the alternate flight plan in NAV mode. When NAV mode is engaged, the flight crew can only clear or modify the TO waypoint by using the DIR key on the MCDU console.

Predictions

The system calculates and displays predictions for all waypoints. It uses the current wind to compute TO waypoint predictions, and uses predicted winds to compute all others.

Constraints

The database may define an altitude and speed constraint for each waypoint of the climb, descent, and approach phases, or the pilot may manually insert such constraints (except at origin, destination, FROM, and pseudo-waypoints). The constraints are displayed in magenta, as long as predictions are not completed. Once predictions are available, constraints are replaced by speed and altitude predictions, preceded by stars. If the star is in magenta, the system predicts that the aircraft will match the constraint (altitude within 250 ft, speed not more than 10 kt above the constraints). If the star is in amber, the system predicts that the aircraft will miss the constraint and the MCDU displays: "SPD ERROR AT WPT".

F-PLN Page B:

This page displays fuel predictions and forecast winds at each waypoint. The pilot calls it up by pressing the NEXT PAGE key when the FLIGHT PLAN A page is displayed.

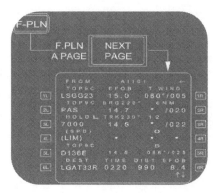

WPT/EFOB/WIND: These lines display consecutive waypoints and associated fuel predictions, and the forecast wind profile. The direction of forecast winds is relative to true north.

Forecast winds include winds entered by the pilot (large font) and the propagated winds at intermediate waypoints (small font). If the flight crew uses a trip wind, it will be displayed for each waypoint. If no other wind entry is made after takeoff, the FROM waypoint will display the actually recorded wind, and the waypoints downpath will still display the trip wind.

Lateral Revision Pages

These pages give the pilot a list of the lateral flight plan revisions, which can be used to change the flight plan beyond a selected waypoint. The pilot calls up these pages from the flight plan pages (A or B) by pressing the left key adjacent to the selected waypoint. Different lateral flight plan revisions are available for different waypoints.

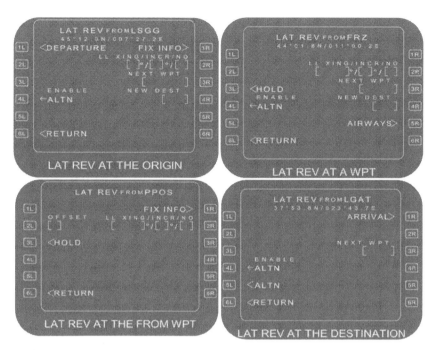

LAT REV AT THE ORIGIN

LAT REV AT A WPT

LAT REV AT THE FROM WPT

LAT REV AT THE DESTINATION

DEPARTURE: This prompt gives the pilot access to the departure pages, where he can select and insert runways, SIDs, and TRANSs.

OFFSET: This prompt allows the pilot to enter a lateral offset, left or right, in the flight plan. The offset may be between 1 and 50 nm. When the pilot enters an offset, the OFFSET field becomes yellow and the 6L and 6R fields display ERASE and INSERT. The pilot can delete an inserted offset either by pressing the CLR pushbutton, by entering a zero for the amount of the offset, or by selecting a DIR TO.

HOLD: This prompt gives access to the hold pages.

ENABLE ALTN: This prompt allows the pilot to switch to the alternate flight plan at the selected revision waypoint, and use it as a new active flight plan. The system never displays this prompt at the FROM waypoint.

ALTN: This prompt gives access to the alternate airport page. The system displays it only at the destination.

ARRIVAL: This prompt calls up the arrival pages, where RWY, APPR, STAR TRANS and VIA can be selected and inserted.

FIX INFO: FIX INFO is only displayed on the lateral revision page at the origin or FROM waypoint. It gives access to the FIX INFO page.

NEXT WPT: The pilot uses this prompt to enter the next waypoint. If this waypoint is a latitude/longitude, or is neither in the database nor in the pilot-defined elements, the display reverts to the NEW WAYPOINT PAGE.

NEXT DEST: The pilot uses this prompt to enter a new destination.

AIRWAYS: The pilot uses this prompt to access the AIRWAYS page.

INSERT: This prompt is displayed when the pilot has created a temporary flight plan. It can be used to activate the temporary flight plan.

Temporary Revision

When the pilot selects a lateral revision, the system creates a "Temporary F-PLN" and displays it in yellow on the MCDU, and as a dashed yellow line on the ND, enabling the pilot to review the data before inserting it. As long as the

36

temporary flight plan is not inserted, the previous flight plan remains active and the system guides the aircraft along it.

Fix Info Page

This page provides access to the RADIAL intercept function. The reference may be one or more radial bearings, based on a given database fix or a pilot-defined element. If the radial intercepts the active flight plan, the intersection point can be converted to a waypoint and inserted into the flight plan. In addition, the ABEAM function may be used. The FIX INFO page may be accessed from the lateral revision page at origin, or at the FROM waypoint.

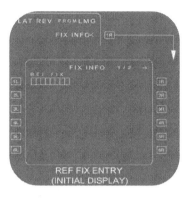

REF FIX: Allows entries of the REF FIX. This reference may be any database element (navaid, waypoint, NDB, airport, runway) or a pilot-defined element. Prior to entry, amber boxes are displayed.

If the radial line intersects the active flight plan, the FMGS will compute the time, the along path DTG (Distance To Go), and the altitude at the intersection point (small green font). A large blue star is then displayed to insert the intersection waypoint into the flight plan. This waypoint is not part of the pilot-stored elements. Format of the created waypoint is: XXXNNN XXX = First 3 letters of REF FIX ident. NNN = Value of the radial

ABEAM: his function enables the pilot to create waypoints on a flight plan (primary or secondary) that are abeam a reference fix. Once computed, the page displays the radial number in large green font. Time, distance and altitude predictions are displayed in small green font. Selecting the key adjacent to the star creates the waypoint and inserts it into the flight plan. The waypoint is identified by AB + the REF FIX ident e.g. AB TLS. Abeam waypoints are not stored in the pilot-stored waypoint database.

Airways Page

This page allows the pilot to select up to five airways for stringing into the flight plan, after the revise waypoint. The pilot calls up this page by pressing the lateral revision page [5R] key.

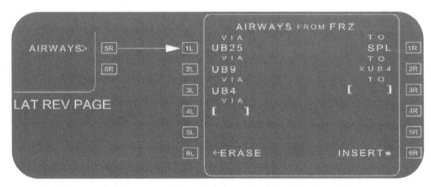

This page displays the airways entered by the pilot. The flight crew presses this key to return to the lateral revision page. This field displays ERASE when a temporary flight plan is created. It enables the temporary flight plan to be erased.

Departure Page

These pages allow the pilot to review departure procedures (RWY, SID, TRANS) and enter them into the active flight plan. When the display shows the lateral revision page for the origin, the pilot calls them up by pressing the 1L key. Three pages are available: RWY, and SIDS and TRANS (if any). The pilot sequentially calls up each page by selecting a data item (such as RWY), or by pressing the NEXT PAGE key on the MCDU console.

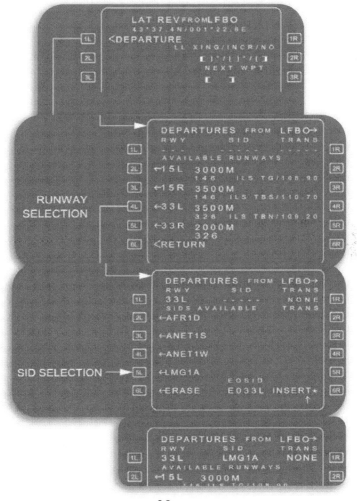

Hold Page

These pages allow the pilot to review and modify the holding pattern parameters at the selected revise waypoint. The flight crew calls up these pages by pressing the HOLD key on the LAT REV page for the waypoint. The flight crew can insert database hold, holds computed by the FMS or holds that they manually define.

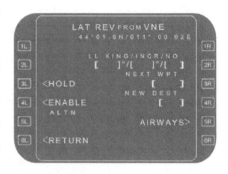

If a hold is defined in the navigation database for the revised waypoint, and can be inserted, the parameters in [1L], [2L] and [3L] appear in yellow. If a default hold is computed by the FMS and can be inserted, the parameters in [1L], [2L] and [3L] appear in yellow

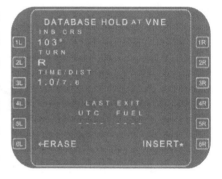

LAST EXIT: This field displays the time at which the aircraft must leave the holding pattern in order to meet fuel policy criteria (extra fuel = 0). The system also displays the estimated fuel on board at that time. Always displayed in thousand of kilograms or pounds.

Arrival Page

Three pages, APPR, STAR, and VIA, are available, along with a fourth, TRANS, if there are any transitions. The pilot calls up each page sequentially, either by selecting a data item (such as APPR), or by pressing the NEXT PAGE key on the MCDU console. The first lines display the APPR, VIA, STAR, and TRANS in green, if they have been inserted in the flight plan, and in yellow, as a temporary flight plan, if they have been selected but not yet inserted. It displays dashes or NONE, if nothing has been selected or inserted.

The pilot presses the key APPR VIA to call up transitions from the last point of the STAR to the first point of the approach. Once the pilot has selected an APPR, STAR, or VIA, the arrow disappears. After the APPR, STAR, or VIA is inserted into the flight plan, it is displayed in green. For each approach, the display shows runway length, heading, and the frequency and identifier of the ILS when ILS is available.

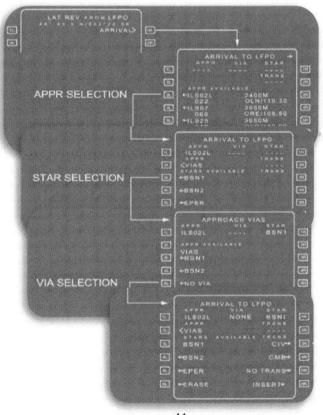

Alternate Page

This page enables the pilot to review, in the NAV database, the alternate airports that are paired with the destination, and define additional alternates, if needed. (Alternate airports are linked to the destination). The pilot calls up this page with the ALTN prompt, from the lateral revision page for the destination.

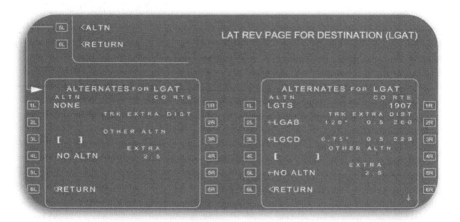

The field ALTN displays the selected alternate: In green, if it is active; in yellow, if it is temporary. "NONE" is displayed, if NO ALTN option is selected, or if the destination has no alternate.

Vertical Revision Pages

These pages contain the menu of available vertical flight plan revisions that can be applied at a selected waypoint. The pilot calls up these pages from the flight plan A or B pages by pressing the right hand key next to the selected revised waypoint. The pilot may make several different vertical revisions (although some may not be available at all waypoints): Speed limit, speed constraint, altitude constraint, time constraint and wind page.

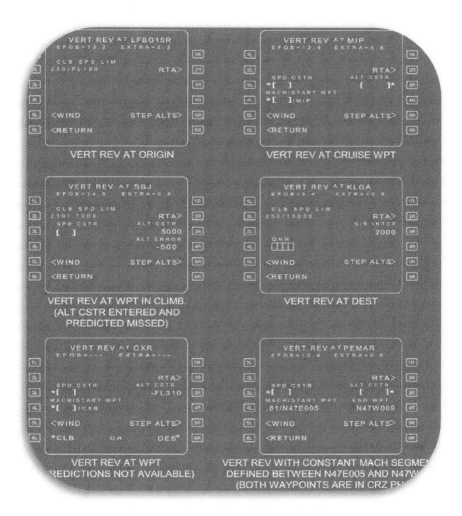

VERT REV AT ORIGIN

VERT REV AT CRUISE WPT

VERT REV AT WPT IN CLIMB.
(ALT CSTR ENTERED AND
PREDICTED MISSED)

VERT REV AT DEST

VERT REV AT WPT
PREDICTIONS NOT AVAILABLE)

VERT REV WITH CONSTANT MACH SEGMEN
DEFINED BETWEEN N47E005 AND N47W
(BOTH WAYPOINTS ARE IN CRZ PH

CLB/DES SPD LIM: This field displays the speed limit applicable to the climb or descent phase. It displays it in large font when data has been inserted manually, and in small font when data comes from the database.

SPD CSTR: This field displays any speed constraint assigned to the revised waypoint. It is in large font when inserted manually, and in small font when it comes from the database.

It is not displayed at the origin airport, at a FROM waypoint, a speed limit pseudo waypoint, or the destination airport.

ALT CSTR: This field displays the altitude constraint assigned to this revised waypoint. It uses large font when the constraint is manually-entered, and small font when it is from the database.

The constraint may be:

"At", entered as XXXXX (Example: FL 180).
"At or above", entered as + XXXXX or XXXXX + (Example: FL + 310). "At or below", entered as – XXXXX or XXXXX – (Example: -5 000).
The altitude window consists of two altitudes between which the aircraft should fly. The crew cannot manually enter a "window" constraint.

DESC: When this field displays "DES", pressing this key assigns the constraints to the descent phase and inserts them into the vertical flight plan. The page reverts to the F-PLN page.

RTA: This prompt gives access to the RTA page. It is not displayed when the VERT REV page is accessed from the alternate F-PLN.

RTA Page

The Required Time of Arrival (RTA) page allows the entry and display of a waypoint identifier, with associated time constraints. The page also displays the entered or computed Estimated Takeoff Time (ETT), as well as the following data:

○   Predicted ETA at the time-constrained waypoint;

○   Performance adjusted SPD target;

o Time error;

o Distance to time constrained waypoint;

o Active speed mode;

The pilot calls up this page with the RTA prompt from the vertical revision page.

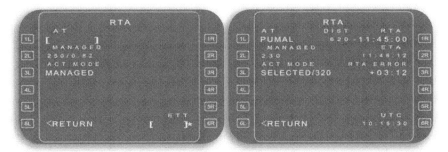

The line displays AT and blue brackets, if no time constraints exist, or AT, DIST and RTA when a time constraint has been defined.
The waypoint identifier is displayed in large blue font.
If only the waypoint identifier has been defined, blue brackets and a blue star are displayed facing the 1R prompt.

ETT: The Estimated Takeoff Time (ETT) field is available in the preflight phase. If no ETT is available, the 6R field displays blue brackets and a blue star. Once available, the ETT is displayed in magenta.

MCDU DIRECT TO Page

Pressing the "DIR" key under the MCDU screen brings up the DIR TO page. On this page, the [1L] key is the DIR TO key. The pilot presses it to modify the flight plan by creating a direct leg from the aircraft's present position to any selected waypoint. When in NAV mode, the pilot must use this key to modify the active leg or the TO

45

waypoint. The pilot cannot call up this page when the aircraft's present position is not valid.

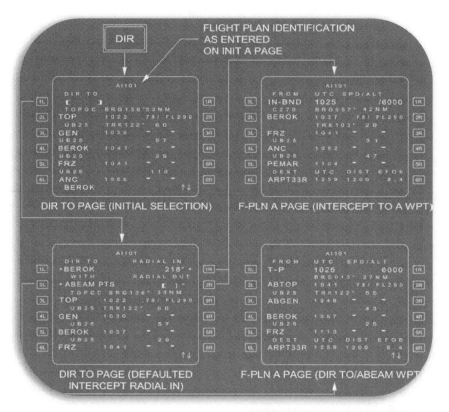

Pressing this key selects the DIRECT TO or INTERCEPT waypoint. The pilot can identify the waypoint to be inserted by using its identifier, its latitude and longitude, place/bearing/distance, or place-bearing/place-bearing. If the pilot

does not select the RADIAL IN (1R) or RADIAL OUT (2R) or ABEAM PTS (2L), the DIR TO function routes the aircraft from the present position to the waypoint inserted in the DIR TO field.

The flight crew presses the key to activate the DIR TO/ABEAM function which projects the flight plan waypoints perpendicularly on the DIR TO leg:

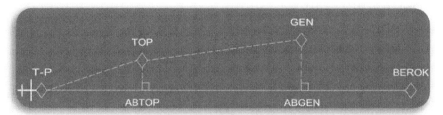

Radial IN/Radial OUT: The pilot fills in these fields to define a radial, associated to the waypoint defined in 1L. These keys respectively activate the DIR TO/INTERCEPT TO and DIR TO/ INTERCEPT FROM functions. The pilot enters the radial in, or radial out, as : XXX, XXX being the radial. The aircraft intercepts from its current position and tracks the selected waypoint and radial to (or from) this waypoint.

If the DIR TO/INTCPT WPT entry is to a waypoint already in the flight plan, a default RADIAL IN is displayed in small font. However, no radial is displayed on the ND for this default radial. No default radial is provided for the RADIAL OUT field.

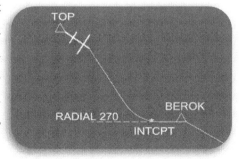

Selecting the INTCPT TO (RADIAL IN [1R]) function: Activates the intercept radial INTO the WPT. Sets the course = radial IN + 180 °. Reverts the display to the F-PLN A page. Selecting the INTCPT FROM (RADIAL OUT [2R]) function: Activates the intercept radial FROM the WPT. Sets the course = radial OUT. Reverts the display to F-PLN A page.

MCDU DATA INDEX Page

There are two INDEX pages. The DATA INDEX 1/2 page gives access to various pages devoted to navigation. The DATA INDEX 2/2 page lists the navigation data, entered in the FMGS. The pilot enters those items labeled "stored" and can modify them. The pilot can call up the others, but cannot modify them. The pilot calls up these pages by pressing the DATA key on the MCDU console:

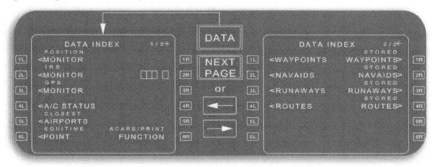

MONITOR: When the flight crew presses these keys, the display shows all essential navigation data.

WAYPOINT: The pilot can call up this page by pressing the 1L key on the DATA INDEX page. The display then shows waypoint information associated with the identifier the flight crew inserts it in the [1L] field. With this page it is possible to call any waypoint not stored in the stored waypoint list, if they belong to the active, temporary, or secondary flight plan.

48

STORED WAYPOINT: The pilot calls up this page by pressing the 1R key on the DATA INDEX page. This page displays waypoints, defined and stored by the pilot. It lists each stored waypoint,

along with a number that shows the relative order in which it was inserted in the database. This number is displayed in the upper righthand corner of the page. For example, "1/20" indicates that the waypoint was the first of 20 stored.

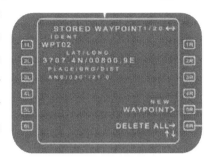

NEW WAYPOINT: The pilot calls up this page by pressing the 5R key on the STORED WAYPOINT page. The pilot can use this page to define and store up to 20 waypoints. Entering an additional waypoint deletes the first one. The pilot defines a waypoint by entering its ident in the data field next to 1L, then by entering its position in the amber boxes. The STORE prompt appears next to 6R when the boxes are filled in, and the pilot presses the key to store the waypoint in the database. If the pilot enters the waypoint's position as place/bearing/distance, or place-bearing/place-bearing, the FMGC computes its latitude and longitude.

NAVAID: The pilot calls up this page by pressing the 2L key on the DATA INDEX page 2. This page displays NAVAID information associated with the identifier the pilot inserts in the [1L] field.

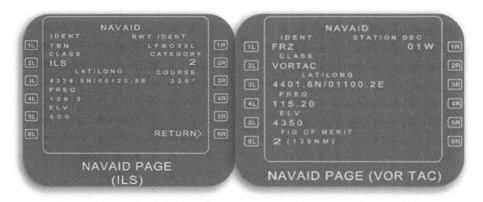

NAVAID PAGE (ILS)

NAVAID PAGE (VOR TAC)

RUNWAY: The pilot uses this page to display or delete the defined and stored runways. The stored runways are listed and numbered in the order in which they were inserted. The number is displayed in the upper righthand corner of the page. (For example, 2/4 means the runway is the second of the four stored runways). The pilot can delete any stored runway from the database by displaying its IDENT in the 1L field, then by pressing the CLR key on the MCDU control panel.

ROUTE: This page displays up to 5 routes, stored by the pilot. The stored routes are listed and numbered in the order of insertion. The number is displayed in the upper right-hand corner of the page.

CLOSEST AIRPORT: The system automatically selects the closest 4 airports from the current aircraft position, and displays them on these pages. A fifth one can be selected by the pilot. Page 1 displays the bearing, distance, and time to go to each airport; page 2 displays the EFOB and allows the crew to enter an effective wind to be flown to each airport. The flight crew accesses the CLOSEST AIRPORTS page 1 by pressing the 5L key from the DATA INDEX A page. They access the CLOSEST AIRPORTS page 2 by pressing the EFOB/WIND prompt (6R key) on page 1.

EQUI-TIME POINT: The pilot uses this page to require an equitime point computation between two different points (airport, NAVAID, runway, NDB or waypoint). This pseudo-waypoint (ETP) is displayed on the navigation display along the F-PLN. The EQUI-TIME POINT page is accessed by pressing the 6L key from the DATA INDEX page:

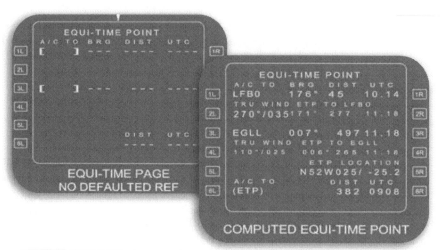

COMPUTED EQUI-TIME POINT

MCDU PERF Page

The flight plan is divided into the following phases: PREFLIGHT, TAKEOFF, CLIMB, CRUISE, DESCENT, APPROACH, GO-AROUND, DONE. Each phase, except for the preflight and done phases, has a performance (PERF) page. The PERF pages display performance data, speeds related to the various phases, and predictions. Pressing the PERF key on the MCDU console calls up the performance page for the current active phase. Performance pages, relating to phases already flown, are not available. In the preflight and done phases, pressing the PERF key brings up the takeoff performance page. Pressing the PERF key in the done phase makes the phase transition to the preflight phase.

PERF TAKEOFF Page: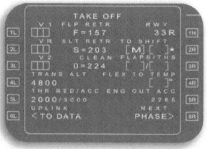

During the preflight phase, the pilot presses the PERF key to call up the takeoff performance page. TAKE OFF is in large white font, when the takeoff phase is not active, and in large green font when it is.

TRANS ALT: This field displays the navigation database default altitude (if defined) once the origin airport is entered. The pilot can modify it.

THR RED/ACC: This is the altitude at which the pilot should reduce the thrust from TOGA/FLX to MAX CLIMB (CL detent) with all engines operative. The thrust reduction altitude defaults to 1 500 ft above the runway elevation, or to the altitude set by the airline
The pilot can modify this altitude: The minimum is 400 ft above the runway elevation. **ACC:** This is the altitude at which the climb phase is triggered. The target speed jumps to the initial climb speed. The default value is 1 500 ft above runway elevation. The flight crew can modify the value. The minimum value is 400 ft above runway elevation, though it is always higher than, or equal to, THR RED.

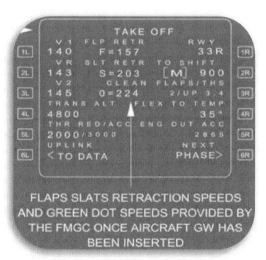

TO SHIFT: The takeoff shift is the distance in meters or feet between the beginning of the runway and the aircraft's takeoff position. When taking off from an intersection, the flight crew should insert this value to ensure a correct update of the FM position. The takeoff shift value must be positive, and cannot be greater than the runway length.

FLAPS/THS: This is an optional flight crew entry for the positions of the flaps and the trimmable horizontal stabilizer (THS) at takeoff. It is for information only (no action). The flight crew can modify it until takeoff, by entering "UP X.X" or "X.X UP", or "DN X.X" or "X.X DN" for the THS.

FLEX TO TEMP: The flight crew inserts the FLX TO temperature for FLX takeoff setting purposes. The flight crew can only enter it during preflight. The system sends it to the FADEC and displays the entered data on the upper ECAM display. The TEMP value is always entered in degrees Celsius.

ENG OUT ACC: This field displays the engine-out acceleration altitude, as defined in the database, or is manually entered by the flight crew. This is for display only, as a reminder. It cannot be cleared. The above ACC altitude rules of [5L] apply to this field.

PERF CLB Page: CLB is displayed in large white fonts when the climb phase is inactive, and in large green fonts if it is active. The field ACT MODE displays the preselected or active speed mode: SELECTED or MANAGED. The pilot cannot modify it from this field. The field CI displays the cost index, as initialized on the INIT A or defaulted from the database, or inserted in this field by the pilot. EO LRC automatically replaces the cost index value in case of engine-out.

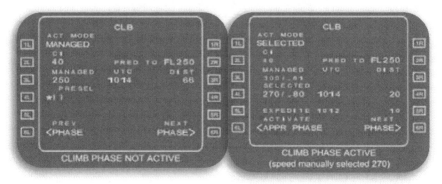

EXPEDITE: This field is blank as long as the aircraft is in preflight. This field displays this legend when the takeoff or climb phase is active. The flight crew cannot engage EXPEDITE from this field. It indicates the time and distance required to reach the altitude displayed in the 2R field, in case of climb at green dot.

ACTIVATE APPR PHASE: The field displays this legend if the climb phase is active. Pressing this key once displays "CONFIRM APPR PHASE*". Pressing it again activates the approach phase.

PERF CRZ Page: in white large font, when cruise phase is not active, in green large font, when it is. The field ATC MODE shows the preselected or active speed mode: SELECTED or MANAGED. The pilot cannot modify it through this field. The field CI shows the cost index as initialized on the INIT A page or defaulted from the database, or as inserted in this field by the crew. EO LRC replaces automatically the cost index value in case of engine out.

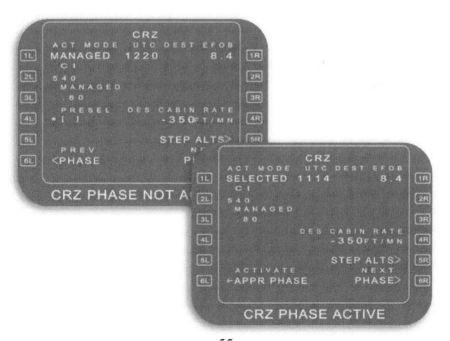

TIME/UTC DES EFOB: Before takeoff this field displays the flight time to destination and the predicted remaining fuel on board. If the crew enters an estimated takeoff time, the field displays automatically the predicted arrival time (UTC) at destination. After takeoff it displays the predicted arrival time at destination (UTC) and the remaining fuel on board. EO CLR is displayed when an engine out is detected.

DES CABIN RATE: This field displays MAX computed DES cabin rate, maximum descent cabin rate. The pilot may modify the value: the FM recomputes then the top of descent in order to match this value. If the FM cannot match the pilot entry, the FM computed value overwrites the pilot entry. A clear action reverts to the default value (-350 ft/min). DES CAB RATE being a negative value, 'minus" is not a necessary entry.

PERF DESC Page: DES is in large white font if the descent phase is not active; it is in large green font, if it is.

EXPEDITE: Displays this legend if the descent phase is active. It indicates the time and distance required to reach the altitude displayed in the 2R field at MMO/VMO speed. The pilot cannot select the EXPEDITE mode through this field.

TIME/UTC DEST EFOB: Before takeoff, this field displays the flight time to destination and the predicted remaining fuel on board. If the crew enters an estimated takeoff time, the field displays automatically the predicted arrival time (UTC) at destination. After takeoff, it displays the predicted arrival time at destination (UTC) and the remaining fuel on board. EO CLR is displayed when an engine-out is detected.

DES PHASE NOT ACTIVE WITH
MANAGED SPEED/MACH SELECTE

DES PHASE ACTIVE WITH
SELECTED SPEED/MACH

PERF APPR Page: APPR is in large white font, if the approach phase is not active; it is in large green font, if it is. The field QNH displays brackets, when the aircraft is more than 180 nm from the destination. Inside 180 nm, a mandatory amber box appears. The flight crew must enter the QNH in hPa or in inches of mercury. The flight crew can modify this entry at any time. The Cabin Pressure Controller (CPC) uses the QNH to compute the cabin pressurization segment. Therefore, an erroneous QNH entry may result in a cabin pressurization that is not appropriate.

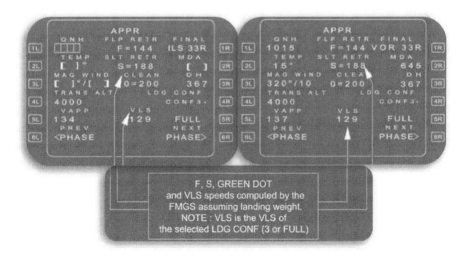

F, S, GREEN DOT
and VLS speeds computed by the
FMGS assuming landing weight.
NOTE : VLS is the VLS of
the selected LDG CONF (3 or FULL)

MAG WIND: The flight crew enters the magnetic wind in knots at the destination in this field. The system transmits any entry made in this field to the vertical revision and flight plan B pages (which display wind direction as true, not magnetic).

TRANS ALT: This field displays the transition altitude taken from the data base (small font) or entered by the flight crew (large font). The flight crew can modify it at any time.

VAPP: The FMGC computes this approach speed, using the formula VAPP = VLS + 1/3 of the headwind components (limited to VLS + 5 as a minimum and VLS + 15 as a maximum). The flight crew can modify VAPP. A clear action reverts VAPP to the computed value.

FINAL: This field displays the approach specified in the flight plan. The flight crew cannot modify it through this field.

MDA/MDH: if the QFE pin program is activated), with associated brackets. The flight crew inserts the value, which it can modify at any time. If the flight crew makes an entry in [3R] or changes the approach, it clears this figure.

DH: If the flight plan includes an ILS approach, this field displays "DH" and empty brackets. The flight crew inserts the decision height. The system will accept an entry of "NO". If the flight crew inserts an MDA or an MDH, this erases the decision height, and this field reverts to brackets. The DH range is 0 to 700 ft.

LDG CONF: The flight crew can select configuration 3 by pressing the 4R key. This moves the * down to the [5R] field, which is displaying "FULL". The flight crew can use this key to select configuration FULL when necessary configuration FULL is the default landing configuration.

PERF GO AROUND Page: GO AROUND is in large white font, if the go-around phase is not active; it is in large green font, if it is.

THR RED ACC: This field displays the thrust reduction altitude and the acceleration altitude.

Thrust reduction altitude:

* Altitude at which thrust must be reduced from takeoff/go-around thrust to maximum climb thrust

* "CLB" or "LVR CLB" flashing on flight mode annunciator

* Defaults to 1 500 ft above destination runway elevation, or to the altitude set by the airline

* Can be modified by the crew (minimum 400 ft above destination runway elevation).

Acceleration altitude:

* Altitude at which target speed jumps to green-dot speed (see the note below)

* Defaults to 1 500 ft above destination runway elevation, or to the altitude set by the airline.

* Can be modified by the crew, but is always equal to (or higher than) the thrust reduction altitude.

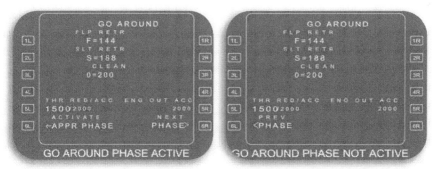

59

MCDU PROG Page

The progress page is a multifunction page that enables the pilot to:

- Select a new cruise flight level

- Crosscheck the navigation accuracy of the Flight Management (FM) system and validate it

- Update the FM position

- Monitor the descent.

The title is different for each flight phase. The vertical phase is in large green font. The flight number is in large white font. EO is large amber font, if the engine-out condition is detected.

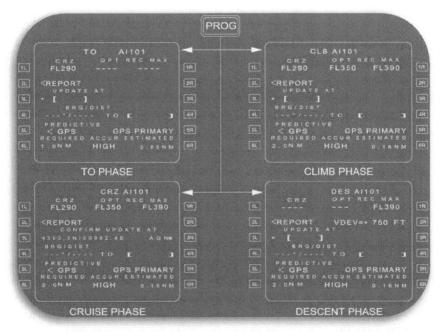

CRZ: This line displays the cruise flight level, inserted on the INIT A page or directly in this field in blue. If the flight crew uses the FCU to select an altitude that is higher than the one displayed in this field, the system changes the number displayed to agree. In this line, the flight crew cannot insert a flight level that is lower than the FCU-selected altitude. This field shows dashes when the descent or approach phase is active.

OPT: This field displays the optimum flight level (in green), that is computed based on the current gross weight, cost index, temperature, wind and a minimum estimated cruising time of 15 min. It displays dashes if an engine-out is detected.

REC MAX: This field displays the recommended maximum altitude (in magenta), that is computed based on the current gross weight and temperature, and assuming that the anti-ice is OFF, if icing conditions are expected). It provides the aircraft with a 0.3 g buffet margin, a minimum rate of climb at MAX CL thrust, and level flight at MAX CRZ thrust. This field is limited to FL 398. If one engine is out, this field displays the recommended maximum engine-out altitude, that is computed based on the long-range cruise speed and assuming that anti-ice is off.

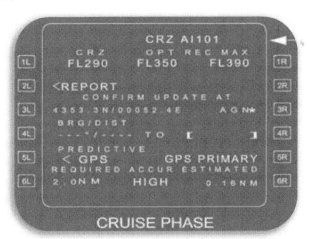

BRG/DIST: On this line, the pilot can enter an airport, a waypoint, a NAVAID, or a runway. The pilot may enter each as an IDENT, a latitude/longitude (LL), a place/bearing/distance (PBD), or a place-bearing/place-bearing (PBX). The field then shows the FMGC-computed bearing and distance from this site to the aircraft's present position. The last distance digit is in 1/10 of a NM. If it does not have an IDENT, the point is called "ENTRY". Example: BRG/DIST 340 °/ 95.4 to ENTRY

MCDU RAD NAV Page

This page enables the pilot to select or verify the radio NAVAIDs, tuned for display purposes only. These NAVAIDs include: VOR, VOR/DME, TAC, VORTAC, ILS, and ADF. If either RMP is set on NAV, this page is blanked on both MCDUs.

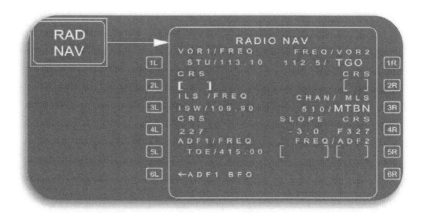

VOR/FREQ: This line displays the identifiers and frequencies of VORs 1 and 2, whether they are automatically or manually tuned. To manually tune a VOR, the pilot inserts the IDENT or frequency. If the IDENT is not in the database, the new NAVAID page comes up. A "clear" action reverts the selection to the autotuned NAVAID.

CRS: This line displays courses for the NAVAIDs in Line 1. The pilot can manually enter the courses through these fields.

ILS/FREQ: This field displays the IDENT and frequency of an ILS. It is autotuned, if the ILS is associated with the departure runway, or if the flight plan shows an ILS approach selected for the destination.

The ILS may also be entered manually. When the manually-entered ILS differs from the ILS that would be autotuned, RWY-ILS MISMATCH appears.

CRS: This field displays the course associated with the ILS in Line 3. It comes up automatically, when an ILS is autotuned, or if an ILS has been manually tuned via its IDENT. Otherwise, the course must be entered manually. The course may be back beam (Bxxx) or front beam (Fxxx).

ADF/FREQ: This line displays the identifiers and frequencies of ADFs 1 and 2. The pilot can use the IDENT or the frequency to manually tune the ADF.

SECUNDARY Pages

The SEC F-PLN key on the MCDU console allows the flight crew to call up the secondary index page and the secondary flight plan page. The secondary flight plan is generally for a diversion, for predictable runway changes for takeoff or landing, or for training.
There are two types of secondary index pages. The type selected depends on the presence of a secondary flight plan.

COPY ACTIVE: The flight crew presses this key to copy the active flight plan into the secondary flight plan and delete the previous secondary plan.

SEC F-PLN: The flight crew presses this key to call up the secondary flight plan pages.

INIT: This field displays this prompt when the secondary flight plan is not defined as a copy of the active flight plan. Pressing this key calls up the SEC INIT A and B pages.

PERF: The flight crew presses this key to call up the performance pages for the secondary flight plan.

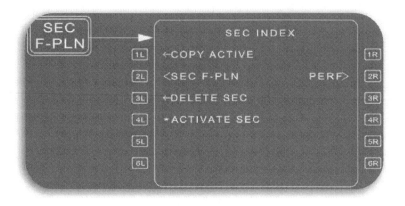

Chapter 2

Domestic Flight I

(MENDOZA to BUENOS AIRES)

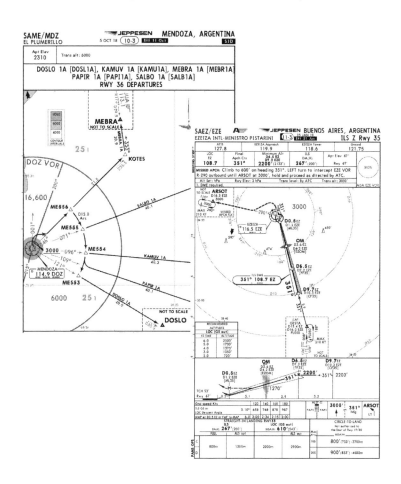

This is for training and entertainment only. For real flight, please see the Airbus manuals.

Domestic flight. Flight information.

In this chapter we will study the work of the Multi Function Control and Display Unit MCDU in a real flight with all the phases involved. It's very important that pilots understand each phase and each step of how the MCDU works. If pilots work with the MCDU correctly, the aircraft will fly correctly, but, if the pilots make mistakes working with the MCDU, the aircraft will follow the wrong instructions.

This flight takes place in Argentina from the city of Mendoza (SAME) to Buenos Aires (SAEZ). The aircraft is an Airbus A320.

Flight Plan:
SAME SID: KAMUV 1A
SAEZ STAR: ASADA 6B
SAEZ APPROACH: ILS RWY35

As a pilot you have to carry out all procedures with the MCDU entering all the information and flight data to complete the cockpit preparation before the flight. After that, the flight will be automatically done until reaching the STAR chart of destination where you will need to operate the MCDU again to carry out the approach procedures.
Let's check the charts before initializing the MCDU

Standard instrumental departure. Mendoza Airport.

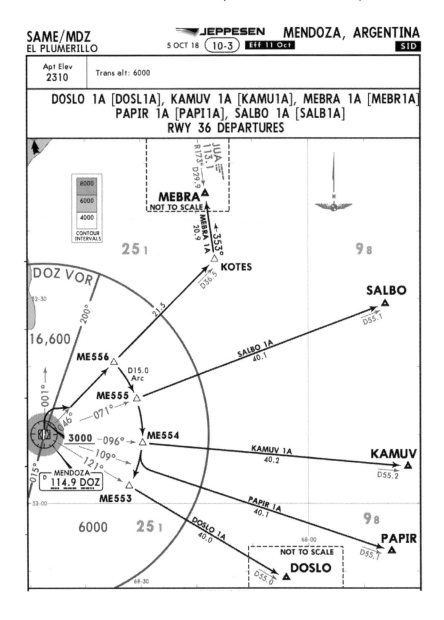

| Apt Elev 2310 | Trans alt: 6000 |

DOSLO 1A [DOSL1A], KAMUV 1A [KAMU1A], MEBRA 1A [MEBR1A]
PAPIR 1A [PAPI1A], SALBO 1A [SALB1A]
RWY 36 DEPARTURES

Standard instrumental arrival. Buenos Aires Airport.

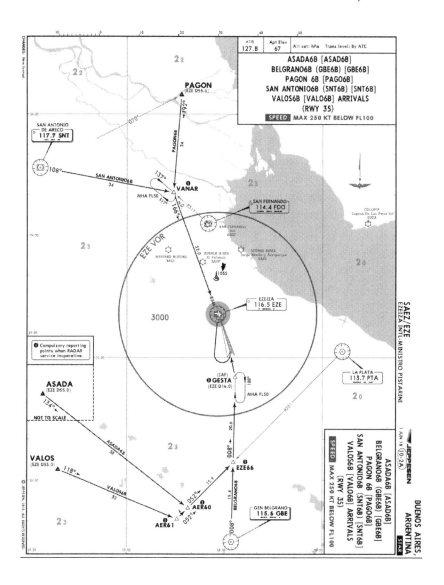

For this example, we will consider the runway 35 in use. The next chart is an ILS Approach from the GESTA position.

69

ATIS	EZEIZA Approach	EZEIZA Tower	Ground
127.8	**119.9**	**118.6**	**121.75**

| LOC EZ **108.7** | Final Apch Crs **351°** | Minimum Alt **D6.5 EZ D7.2 EZE 2200'**(2133') | ILS DA(H) **267'**(200') | Apt Elev 67' Rwy 67' |

MISSED APCH: Climb to 600' on heading 351°. LEFT turn to intercept EZE VOR
R-290 outbound until ARSOT at 3000', hold and proceed as directed by ATC.

Alt Set: hPa	Rwy Elev: 2 hPa	Trans level: By ATC	Trans alt: 3000'

1. DME required. MSA EZE VOR 3000

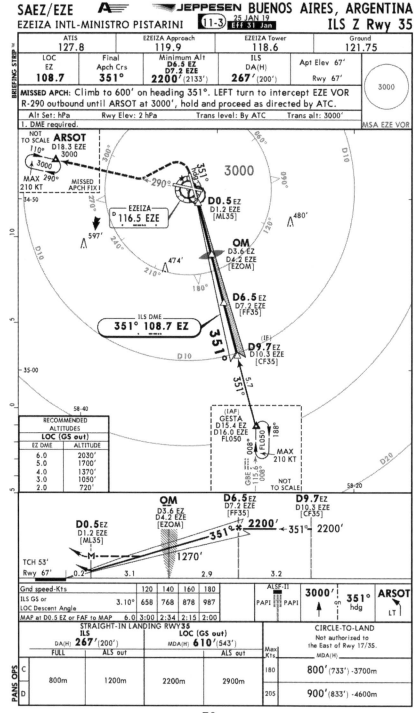

RECOMMENDED ALTITUDES LOC (GS out)	
EZ DME	ALTITUDE
6.0	2030'
5.0	1700'
4.0	1370'
3.0	1050'
2.0	720'

Gnd speed-Kts		120	140	160	180
ILS GS or LOC Descent Angle	3.10°	658	768	878	987
MAP at D0.5 EZ or FAF to MAP	6.0	3:00	2:34	2:15	2:00

ALSF-II PAPI ⊡ PAPI **3000'** on **351°** hdg **ARSOT** LT

STRAIGHT-IN LANDING RWY35					CIRCLE-TO-LAND
ILS DA(H) **267'**(200')		**LOC (GS out)** MDA(H) **610'**(543')			Not authorized to the East of Rwy 17/35.
FULL	ALS out		ALS out	Max Kts	MDA(H)
800m	1200m	2200m	2900m	180	**800'**(733') -3700m
				205	**900'**(833') -4600m

PANS OPS: C, D

70

It's the first flight of the day. The aircraft is in "cold and dark" mode. When the pilot reaches to the cockpit, he turns on the systems. The MCDU is off. To turn it on it is necessary to set the BRT knob on,. It is located on the right side of the MCDU, as you can see in the next picture:

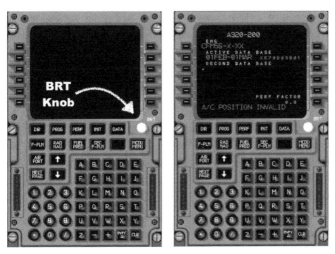

The other systems are on, but in standby mode until the pilot inserts the information into the MCDU. The PFD (primary flight display) does not show information. The ND (navigation display) shows the message MAP NOT AVAIL and the FCU (flight control unit) is ready to set all parameters.

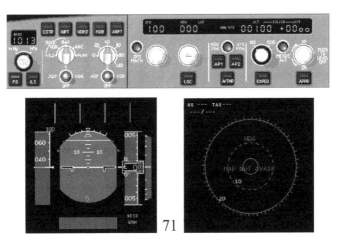

71

The next step is to initialize the system from the **INIT page**. Here, the pilot must insert all information about, departure and arrival airport, route, flight level, temperature, wind, cost index, and flight number. For that purpose, the pilot uses the alphanumeric keys to write each data and the lateral keys to insert them in the correct place, as you can see in this picture,

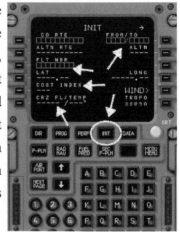

For this example, it's necessary to insert SAME/SAEZ (ICAO codes) in the section of FROM/TO. Each word that the pilot writes appear at the bottom of the display but isn't inserted yet and system remains without information until the pilot inserts it in the correct place.

To clear any wrong word, the pilot has to press the CLR key located at the bottom of the keyboard.

After inserting the flight data, the message ALIGN IRS (inertial systems) appears. This is necessary to set the correct position of the aircraft over the airport. When the IRS system is initialized, the PFD and the ND show the initial information.

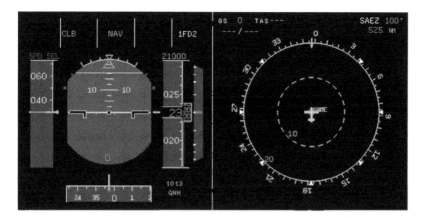

When the INIT page is completed, it shows the information to check and if it's correct, the pilot can continue with the next page. Remember, if there's any error, just write the correct data and insert it again in the correct place. The wrong information will be replaced by the correct information.

The next page is the INIT B. Here the pilot must set the information about weight and fuel. To access to the INIT B page from the INIT A page, just press the NEXT PAGE key. When this setting is ready, the INIT B page shows the value of takeoff weight and landing weight on the right side of the display. In this example TOW 69.3 and LW 30.8.

73

The system is ready. The next step will be to set the navigation from the F-PLN page. This page shows each point of the route, from departure airport to destination airport and every waypoint between them. The Pilot just needs to set the SID (standard instrumental departure), the STAR (standard arrival) and the

approach chart to the runway in use for the destination. This example shows the message F-PLN DISCONTINUITY due to that the SID is not loaded yet. In order to select the SID for the runway in use it is necessary to press the L1 key and then the next page of LATERAL REVISION will show with all the options available.

This new page titled LAT REV from SAME (ICAO code) is to confirm the correct airport. The key L1 on the left side shows the option DEPARTURE to select the runway for takeoff and its SID.

When the pilot presses the DEPARTURE key, two runway options will show, RWY 18 and RWY 36. Beside them, the runway length in meters. And finally, below the runways the information about the ILS.

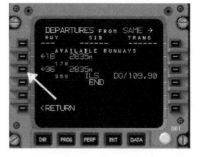

74

In this example, the runway in use is 36 and the SID to use is KAMU3A. The pilot has to select this option and insert it. Before inserting the SID, its track will be shown with a dashed line on the ND to confirm the correct courses.

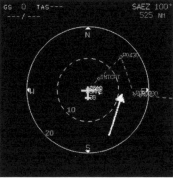

After inserting the SID, its track will be shown with a normal line on the ND, as shown in this picture. Note that the message F-PLAN DISCONTINUITY has disappeared.

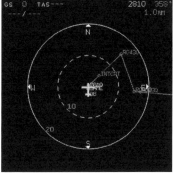

Once everything is loaded, the runway with the correct SID, it's time to select the destination options. To do so, it's necessary to press the last key on the bottom of the display where DEST and SAEZ (ICAO code of destination) are indicated. After pressing this key, the LAT REV page will appear like in the previous step.

When activating this option, the pilot can select not only the Approach procedure for any runway, but also the STAR chart.

After pressing the key, there appears a new window with the option ARRIVAL on the right side indicating LAT REV from SAEZ to confirm the correct airport.

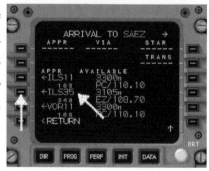

The next page shows the available runways and the IFR procedures for each runway with its data. For this example, the pilot has to select the runway 35 and an ILS procedure. This option is ILS35.

After selecting the desired runway, the STAR CHART options appear. In this case, the correct option is ASAD5B. Before inserting the selection, the ND shows the track to fly with a dashed line. If it's correct, the pilot has to confirm and insert this selection.

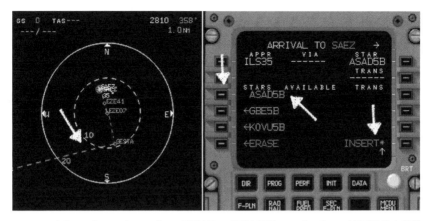

After inserting the STAR CHART and the APPROACH CHART, its track will be shown with a normal line on the ND, as shown in this picture.

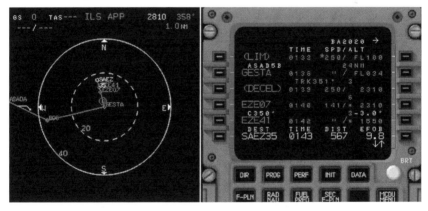

Once the F-PLAN page is completed, the next page will be RAD NAV where the pilot has to set the Nav aids information based on the standard instrument departure chart (SID) to fly. In this case, KAMUV SID. This page gives the option to set two VOR, one ILS and two ADF frequencies, each one with its courses to fly. In this example, the pilot has to select the VOR DOZ frequency with both courses and the ILS frequency in case of any immediate return due to any emergency.

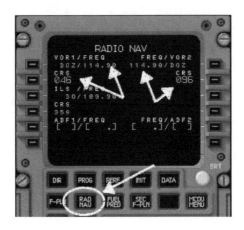

Next page is informative. In the PROG (progress) page, the pilot has to check the optimum and maximum flight level. Additionally, the crew can select any waypoint or any nav aid to obtain the bearing and distance to this reference at any moment during the flight. In this example, the optimum flight level is FL368 and the maximum flight level is FL373. Additionally, the DOZ VOR is selected to obtain bearing and distance.

The next is one of the most important pages. In PERF (performance) page, the pilot has to set speeds, altitudes, flaps setting and the takeoff power.

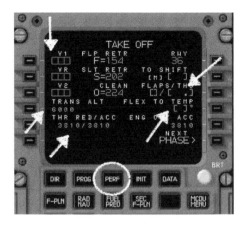

In this example, the pilot has set the speeds (V1:122kt. VR: 125kt. V2:144kt), flaps setting at position 3, and the takeoff power with a value of 55°.

Finally, there is an option that allows the pilot to set a secondary flight plan but in an inactive mode. This option is useful in case of an emergency where by is necessary to change the flight plan immediately.

In this page SEC F-PLN, the system shows three options, copy the active flight plan, create a new one, and set the INIT page for this new flight plan.

When the system is ready and all takeoff data is loaded, the PFD and ND show the information of this phase. The PDF shows information about active speed, altitude, and pitch. The ND will show the track line of the SID to fly.

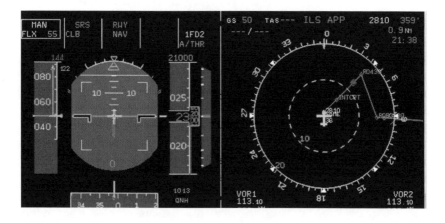

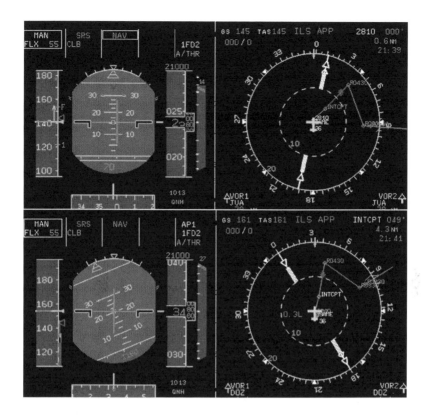

When the pilot changes the ND's view mode, from ROSE NAV to ARC NAV, the display shows a better and more complete illustration of the SID.

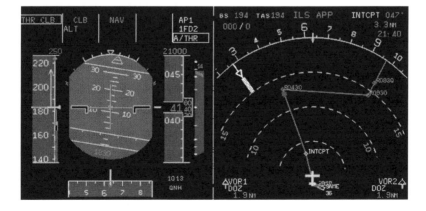

During a normal flight, the aircraft will fly the SID's track as the pilot has planned it. In this example, the air traffic controller will request to fly directly to KAMUV position disregarding the SID.

To carry out this procedure requested by the air traffic controller, the pilot has to change the navigation loaded on the MCDU and take the aircraft directly to KAMUV position. To do so, the pilot has to press the DIR key and then, select the waypoint desired (kamuv) from que lateral keys. The ND will change and show the new direct route to KAMUV without the SID track.

82

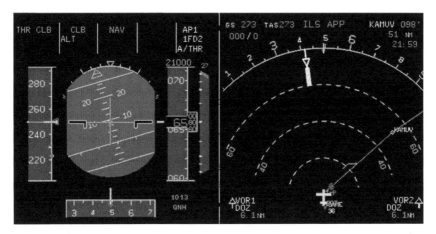

Once the aircraft is flying directly to KAMUV position, on the ND, a blue arrow appears in the middle of the route track. This symbol indicates the TOC or Top of Climb and it appears in the MCDU too but as (T/C) and indicating the distance remaining to this point. At this point, the aircraft will change from the climb phase to the cruise phase.

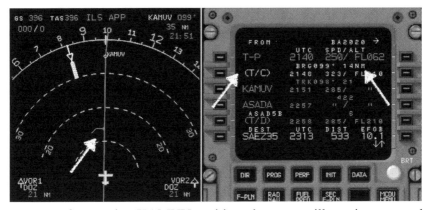

After passing KAMUV position, the route will continue several miles until the next waypoint. For this example, we will assume that the aircraft is near to its destination at 90 NM from the ASADA position, the next waypoint prior to the approach. In the MCDU the message ENTER DEST DATA appears indicating that the pilot must insert the destination data.

Here the pilot must set again the parameters and information on some pages in the MCDU to be ready for initiation of the descent, the approach and landing. The first page is RAD NAV to set the nav aids frequencies. In this example, the frequency of EZE VOR and EZ ILS.

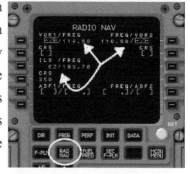

The next page is PROG, where the pilot can set some nav aid or waypoint to get the bearing and distance to this reference. It's not necessary but it's very helpful. In this case, the reference will be EZE VOR.

The aircraft is approaching ASADA position and near to the TOD or Top of Descent. This point is shown with a white arrow in the ND over the route line. In this case, just a few miles after ASADA. At this point of the navigation, the pilot needs to change the flight's phase from cruise to descent and approach phase. To do that, on the PERF page appears the option ACTIVATE APPR PHASE.

When the pilot activates this option, the aircraft will change the phase and the PERF page will change too. Now, on this page are available the destination options such as QNH, TEMP, WIND, DH/DA, FLAPS CONFIGURATION, and more. In this case, qnh 1013, temp 12°, wind 300°/006, dh 200, flaps conf 3, for ILS35 as planned.

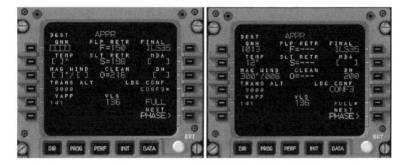

The aircraft is descending and approaching to ADASA position. Here the air traffic controller request to fly direct to the GESTA position, the last waypoint prior to the final approach. To do that, the pilot has to press the DIR key and select GESTA with the lateral keys, like before. The ND will change from the STAR's diagram to a diagram direct to the GESTA.

Unless the pilot sets an overfly point over GESTA, the aircraft will lateral pass to this point. In this case, the air traffic controller will request a holding at GESTA. To carry out this procedure, the pilot has to set a hold on GESTA by pressing the lateral key of this waypoint and selecting the HOLD option.

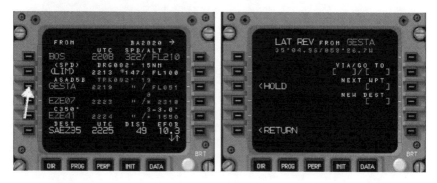

When the pilot inserts the holding data, the ND shows the information to load with dashed lines until the pilot confirms this procedure.

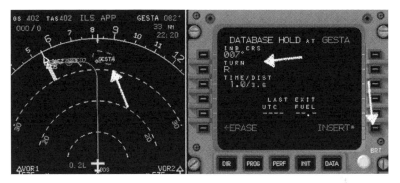

After pressing the INSERT prompt, the ND will show the new route with a hold loaded on GESTA position and this hold will appear as a new point.

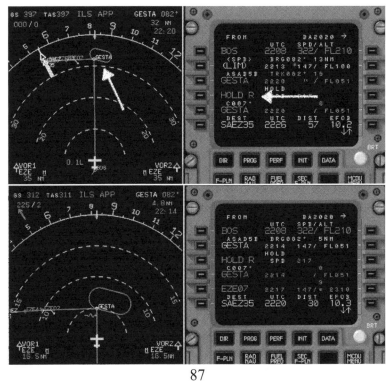

Reaching GESTA position, the air traffic controller will request again another direct flight, in this case direct to the final approach fixed EZE07 prior to continue with the ILS references.

Finally, the aircraft will capture the ILS signal and will fly the lateral navigation with LOC and the vertical navigation with GS until the DH prior to touchdown or go around. This procedure will be flown automatically by the system, but the pilot can take the control at any moment and continue flying manually.

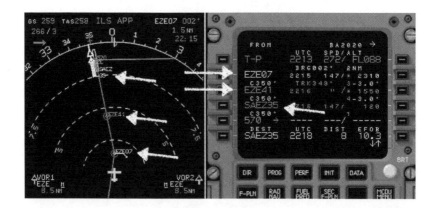

Chapter 3
Domestic Flight II

(MALAGA to VALENCIA)

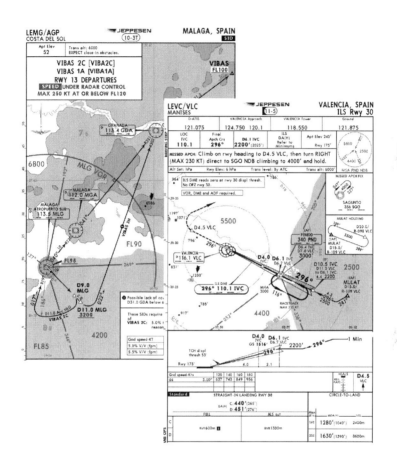

This is for training and entertainment only. For real flight, please see to the Airbus manuals.

Domestic flight II. Flight information.

In this chapter we are going to study the work of the MCDU in a real flight with all the phases involved. It's very important that pilots understand each phase and each step of how the MCDU works. If pilots work with the MCDU correctly, the aircraft will flight correctly, but if the pilots make mistakes working with the MCDU, the aircraft will follow the wrong instructions.

This flight takes place in Spain from Malaga City (LEMG) to Valencia City (LEVC). The aircraft is an Airbus A320. The charts for this flight are:

Flight Plan:
LEMG SID: VIBAS 1A
AIRWAY: UM985
LEVC STAR: ASTRO 1B
LEVC APPROACH: ILS RWY30

As a pilot you have to carry out all procedures with the MCDU inserting all the information and flight data to complete the cockpit preparation before the flight. After that, the flight will be automatically done until reaching the STAR chart of destination where you will need to operate the MCDU again to carry out the approach procedures. In this case, you have to be ready for an immediate return and landing. To plan this situation, it's necessary to set an approach to Malaga again for runway 13 in the secondary flight plan. Let's check the charts before initializing the MCDU.

Standard instrumental departure. Malaga Airport.

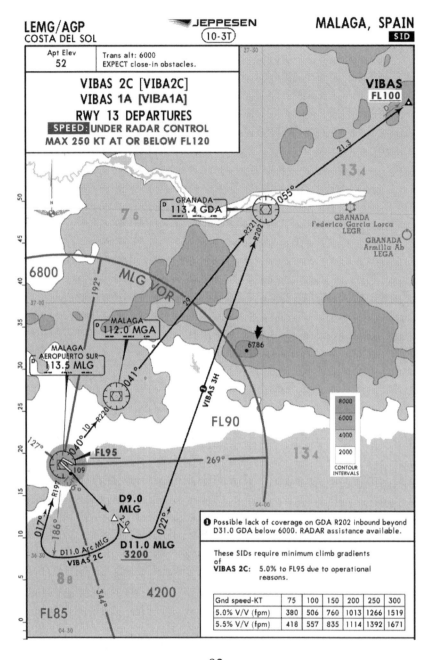

Airway and waypoint. Extract of the enroute's chart.

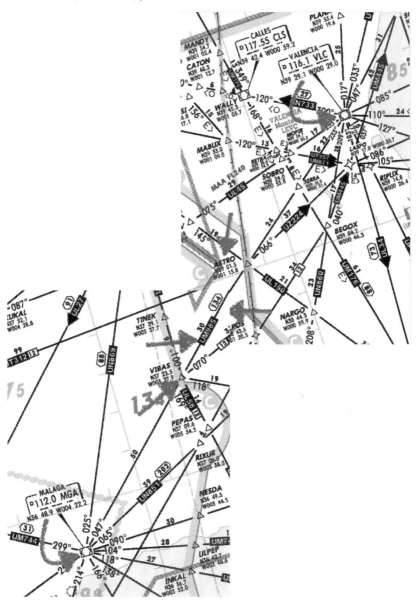

Standard instrumental arrival. Valencia Airport.

ATIS	Apt Elev	Alt Set: hPa
121.075	240	Trans level: By ATC

ASTRO 1C [ASTR1C]
MABUX 2Y [MABU2Y]

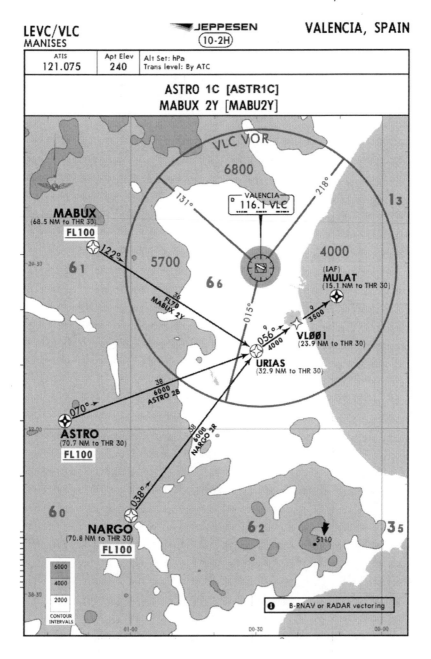

B-RNAV or RADAR vectoring

94

For this example, we will consider the runway 30 in use. The following chart is an ILS Approach from MULAT position:

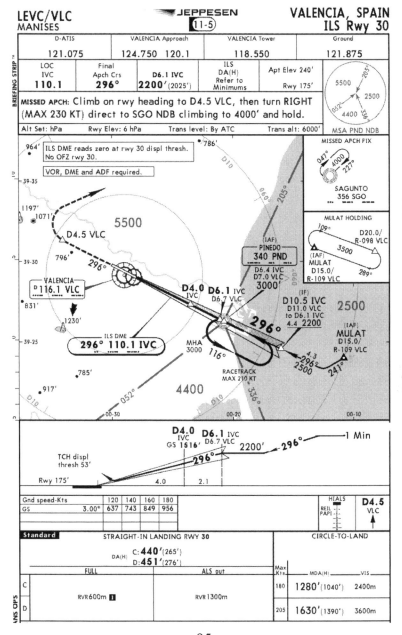

It's the first flight of the day. The aircraft is in "cold and dark" mode. When the pilot gets in the cockpit, he turns the systems on. The MCDU is off. To turn it on it is necessary to set the BRT knob on, which is located on the right side of the MCDU as you can see in the following picture:

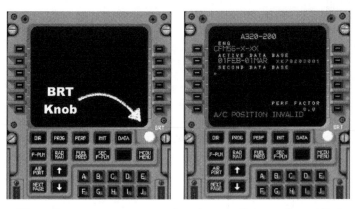

The other systems are on but in standby mode until the pilot inserts the information in the MCDU. The PFD (primary flight display) does not show any information. The ND (navigation display) shows the message MAP NOT AVAIL and the FCU (flight control unit) is ready to set all the parameters.

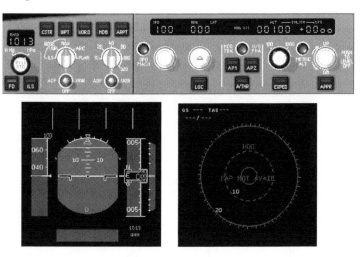

The next step is to initialize the system from the INIT page. Here, the pilot must insert all information about departure and arrival airport, route, flight level, temperature, wind, cost index, and flight number. To do it, the pilot uses the alphanumeric keys to write each data and the lateral keys to insert them in the correct place, as you can see in this picture.

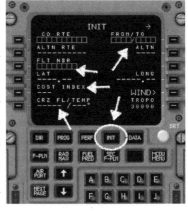

For this example, it's necessary to insert LEMG/LEVC (ICAO codes) in the section of FROM/TO. Each word that the pilot writes appear at the bottom of the display but isn't inserted yet and the system remains without information until the pilot inserts it in the correct place.

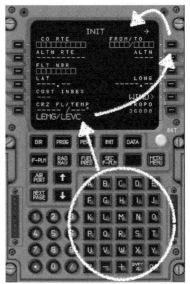

To clear any wrong word, the pilot has to press the CLR key located at the bottom of the keyboard.

After inserting the flight data, the message ALIGN IRS (inertial systems) appears. This is necessary to set the correct position of the aircraft over the airport. When the IRS system is initialized, the PFD and the ND show the initial information. The ND shows the ICAO code

97

and the distance to the destination airport. After setting the flight level on the FCU (flight control unit), the PFD shows this flight level over the altitude column.

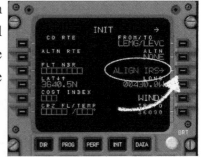

When the INIT page is completed, it shows the information to check and if it's correct, the pilot can continue with the next page. Remember, if there's any error, just write the correct data and insert it again in the correct place. The wrong information will be replaced by the correct information.

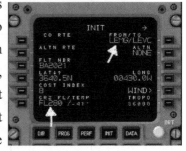

The next page is INIT B. Here the pilot must set the information about weight and fuel. To access to the INIT B page from the INIT A page, just press the NEXT PAGE key. When this configuration is ready (ZFW + FUEL), the INIT B page shows the value of takeoff weight and landing weight in the right side of the display. In this example TOW 73.3 and LW 54.4.

The system is ready. The next step will be to set the navigation from the F-PLN page. This page shows each point of the route, from the departure airport to the destination airport and every waypoint in between. The Pilot just needs to set the SID, the STAR ,and the approach chart to the runway in use for the destination. In this example appears the message FPLN DISCONTINUITY due that the SID is not loaded yet. In order to select the SID for the runway in use it is necessary to press the L1 key and then the next page of LATERAL REVISION with all the options will appear.

In this new page shows the title LAT REV from LEMG (ICAO code) to confirm the correct airport. The key L1 on the left side shows the option DEPARTURE to select the runway for takeoff and its SID.

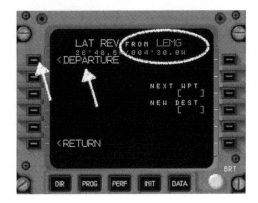

When you press the DEPARTURE key, two runway options will show, RWY 13 and RWY 31. Beside them, the runway length. And finally, below the runways, the information about the ILS.

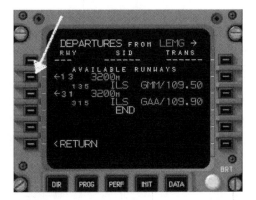

In this example, the runway in use is 13 and the SID to use is VIBAS 1A. The pilot has to select this option and insert it. Before inserting the SID, its track will be shown with a dashed line on the ND to confirm the correct courses.

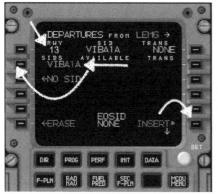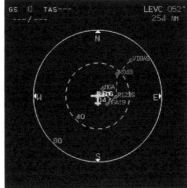

After inserting the SID, its track will be shown with a normal line on the ND, as shown on this picture. Notice that the message FPLAN DISCONTINUITY has disappeared.

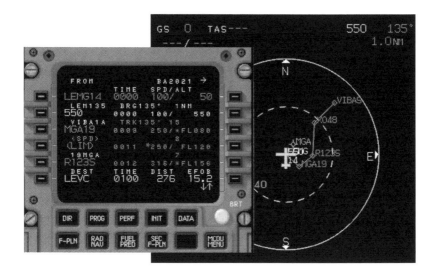

Once everything is loaded, the runway with the correct SID, it's time to select the destination options. To do it you need to press the last key at the bottom of the display where DEST and LEVC (ICAO code of destination) are indicated. After pressing this key, the LAT REV page will show like at the previous the step.

When activating this option, the pilot can select not only the Approach procedure for any runway, but also the STAR chart.

After pressing the key, a new window shows with the option ARRIVAL on the right side, indicating LAT REV from LEVC, to confirm the correct airport.

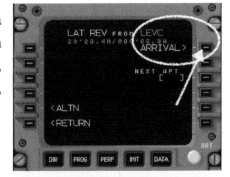

The next page shows the available runways and the IFR procedures for each runway with its data. For this example, the pilot has to select the runway 30 and an ILS procedure. This option is ILS30.

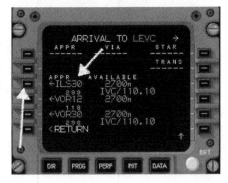

Once the desired runway is selected, the STAR CHART options will show. In this case the correct option is ASTR1C. Before inserting the selection, the ND shows the track to fly with a dashed line. If it's correct, the pilot has to confirm and insert this selection.

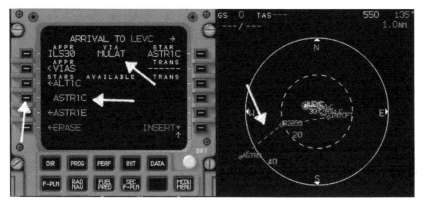

After inserting the STAR CHART and the APPROACH CHART, its track will be shown with a normal line on the ND, as shown on this picture.

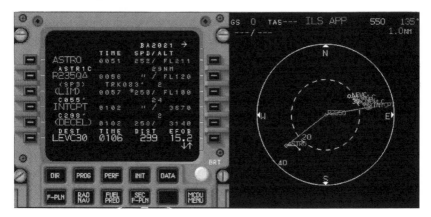

Once the F-PLAN page is completed, the next page will be RAD NAV where the pilot has to set the Nav aids information based on the standard instrument departure chart (SID) to fly. In this case, VIBAS 1A SID. This page gives the option to set two VOR, one ILS and two ADF frequencies, each one with its courses to fly. In this example, the pilot has to select the VOR GDA frequency with both courses and the ILS frequency in case of any immediate return due to any emergency.

The next page is informative. In the PROG (progress) page, the pilot has to check the optimum and maximum flight level. Additionally, the crew can select any waypoint or any nav aid to obtain the bearing and distance to this reference at any moment during the flight. In this example, the optimum flight level is FL290 and the maximum flight level is FL362. Also, the MGA VOR is selected to obtain bearing and distance.

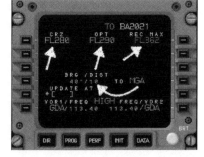

The next is one of the most important pages. In PERF (performance) page, the pilot has to set speeds, altitudes, flaps setting and the takeoff power.

In this example, the pilot has set the speeds (V1:124kt. VR: 130kt. V2:135kt), flaps setting at position 3 and the takeoff power with a value of 65°.

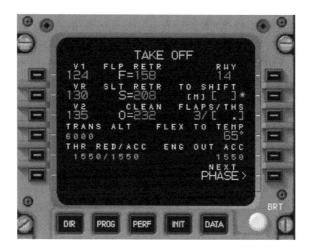

Finally, there is an option that enables the pilot to set a secondary flight plan but in an inactive mode. This option is useful in case of an emergency where it is necessary to change the flight plan immediately. In this page SEC F-PLN, the system shows three options, copy the active flight plan, create a new one, and set the INIT page for this new flight plan. In this case, the pilot needs to set the secondary flight plan with an immediate return just for a case of an emergency, but this plan will remain in standby mode until the pilot activates it, if needed.

To do so, the pilot has to copy the active flight plan, then, in the F-PLN page, he has to change the destination pressing the lateral key after VIBAS position. Here shows a new page with the option NEW DEST on the right side to insert LEMG. This action will change the destination airport and it will be available to select an approach procedure.

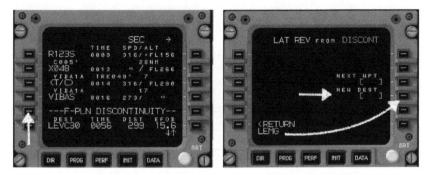

Once the destination is changed, the pilot has to select the runway and its approach procedure by pressing the lateral key and following the options. In this case, ILS for runway 13 via BAMAR from VIBAS position.

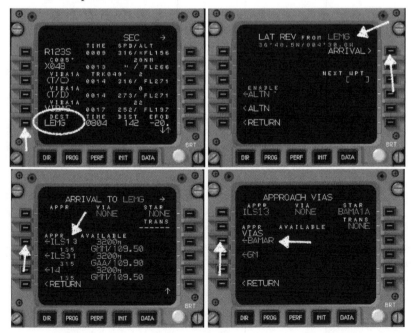

After inserting the ILS procedure, the ND will show the new track from VIBAS position to BAMAR position for an ILS approach to the runway 13.

Finally, the secondary flight plan will remain inactive unless the pilot decides to activate it.

When the system is ready and all takeoff data is loaded, the PFD and ND show the information of this phase. The PDF shows information about active speed, altitude and pitch. The ND will show the track line of the SID to fly.

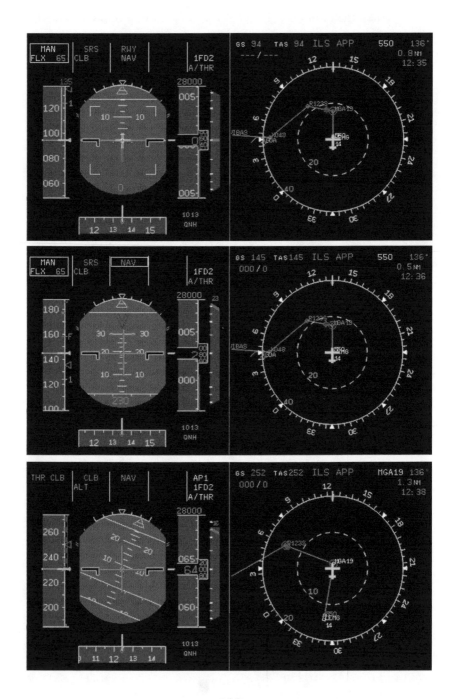

When the pilot changes the ND's view mode, from ROSE NAV to ARC NAV, this display shows a better and more complete illustration of the SID.

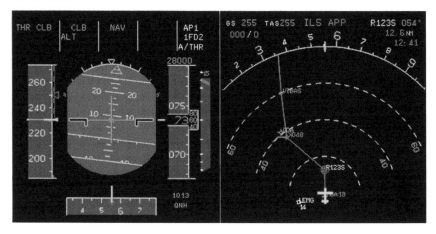

During a normal flight, the aircraft will fly the SID's track as the pilot has planned it. In this example, the air traffic controller will request to fly direct to VIBAS position disregarding the SID.

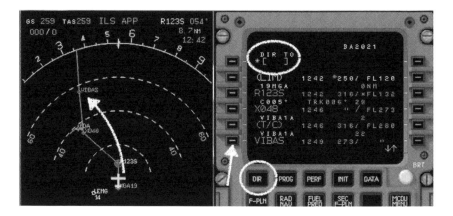

To carry out this procedure requested by the air traffic controller, the pilot has to change the navigation loaded on the MCDU and takes the aircraft direct to VIBAS position. To do so, the pilot has to press the DIR key and then, select the waypoint desired (vibas) from que lateral keys. The ND will change and show the new direct route to KAMUV without the SID track.

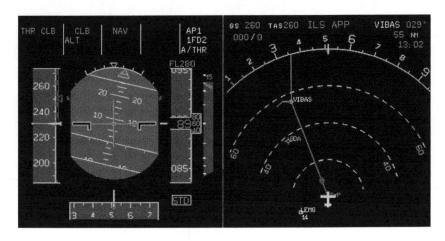

Once the aircraft is flying directly to VIBAS position, on the ND a blue arrow appears in the middle of the route track. This symbol indicates the TOC or Top of Climb and it appears in the MCDU too but as (T/C) and indicating the distance remaining to this point. At this point the aircraft will change from climb phase to cruise phase.

After passing the VIBAS position, the route will remain for several miles until the next waypoint. For this example, we will suppose

110

that the aircraft is near to destination at 91 NM from ASTRO position, the next waypoint prior to approach. In the MCDU the message ENTER DEST DATA appears, indicating that the pilot must insert the destination data.

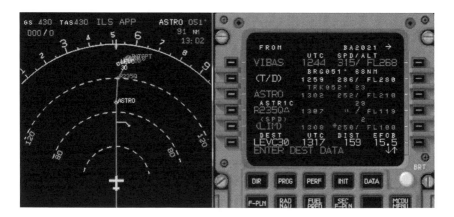

Here the pilot must set again the parameters and information in some pages in the MCDU to be ready for initiation of the descent, the approach, and landing. The first page is RAD NAV to set the nav aids frequencies. In this example, the frequency of VLC VOR and IVC ILS.

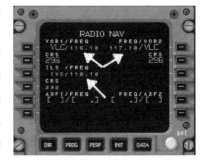

The next page is PROG, where the pilot can set some nav aid or waypoint to get the bearing and distance to this reference. It's not necessary but it's very helpful. In this case, the reference will be VLC VOR.

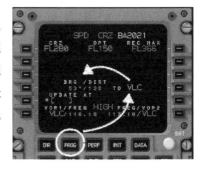

The aircraft is approaching to ASTRO position and near to the TOD or Top of Descent. This point is shown with a white arrow in the ND over the route line. In this case, just a few miles before ASTRO. At this point of the navigation, the pilot needs to change the flight's phase from cruise to descent and approach phase. For that purpose, in the PERF page appears the option ACTIVATE APPR PHASE.

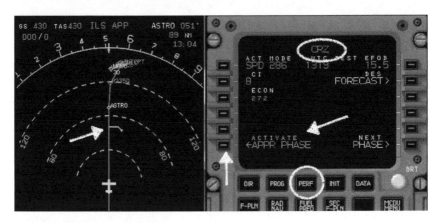

When the pilot activates this option, the aircraft will change the phase and the PERF page will change too. Now in this page are available the destination options such as QNH, TEMP, WIND, DH/DA, FLAPS CONFIGURATION and more. In this case, qnh 1013, temp 12°, wind 010°/005, dh 440, flaps conf 3, for ILS30 as planned.

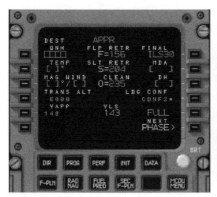
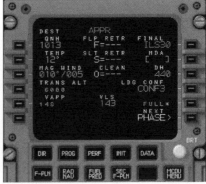

Just in case of any change, the pilot decided to set the secondary flight plan for an approach procedure for the runway 12. This secondary flight plan will remain in standby mode. To do that, the pilot has to copy the active plan and then change the arrival option with the new approach procedure.

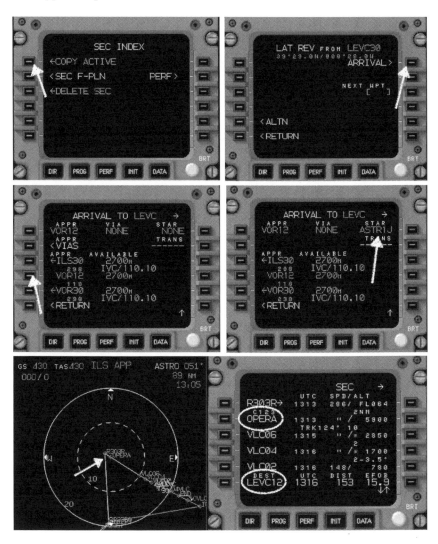

The flight continues normally reaching the ASTRO position and ready for an ILS approach to the runway 30. At this point the air traffic controller requests to change the runway due to traffic.

To proceed as requested by the air traffic controller, the pilot has to activate the secondary flight plan in which the approach procedure is ready for runway 12.

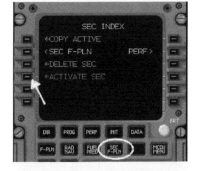

Once the secondary flight plan is activated, on the F-PLN page changes the destination from LEVC30 to LEVC12 and the ND shows the new track to fly.

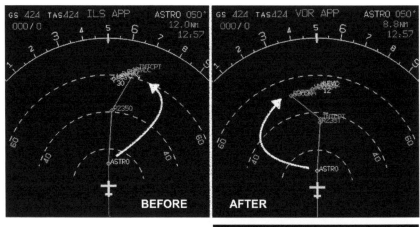

BEFORE | AFTER

After passing the ASTRO position, the air traffic controller requests to fly directly to OPERA position. To do that, the pilot has to press the DIR key and select OPERA with lateral keys, like before. The ND will change the diagram direct to OPERA.

Unless the pilot sets an overfly point over OPERA, the aircraft will pass laterally to this point. In this case, the air traffic controller will request to hold over OPERA position. To carry out this procedure, the pilot has to set a hold over OPERA pressing the lateral key of this waypoint and selecting the HOLD option.

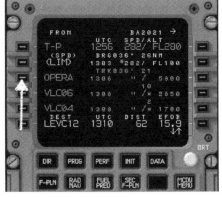

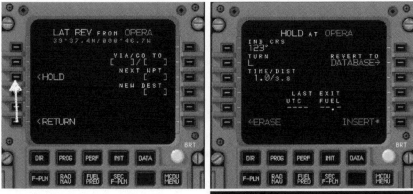

When the pilot inserts the data on hold, the ND shows the information to load with dashed lines until the pilot confirms this procedure.

After pressing the INSERT prompt, the ND will show the new route with a hold loaded over OPERA position and this hold will appear as a new point.

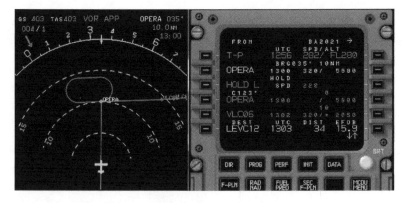

Prior to complete the holding procedure and continuing with the final approach, the pilot has to exit from the holding pattern by pressing the lateral key.

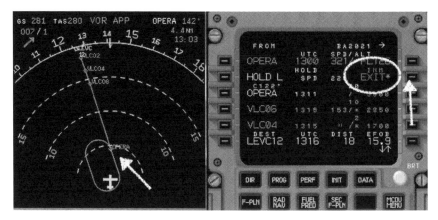

Finally, the aircraft will continue the final approach. This procedure will be flown automatically by the system, but the pilot can take the control at any moment and continue flying manually.

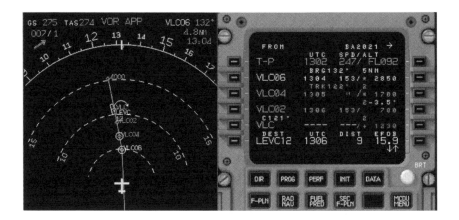

Chapter 4
International Flight I

(SAO PAULO to SANTIAGO DE CHILE)

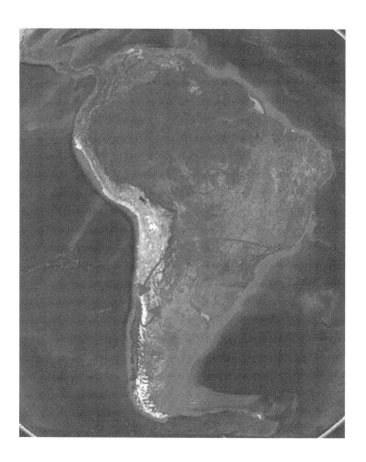

This is for training and entertainment only. For real flight, please see the Airbus manuals.

International flight I. Flight information.

In this chapter we are going to study the work of the MCDU in a real flight with all the phases involved . It's very important that pilots understand each phase and how the MCDU works step by step. If pilots work with the MCDU correctly, the aircraft will flight correctly, but if the pilots make mistakes working with the MCDU, the aircraft will follow the wrong instructions.

This flight takes place in south America from Sao Paulo City (SBGR) to Santiago de Chile City (SCEL). The aircraft is an Airbus A320. The charts for this flight are:

Flight Plan:
SBGR SID: CGO 1
SCEL STAR: UMKAL 4A
SCEL APPROACH: ILS RWY17L
SCEL SID: ALBAL 6 (to the alternative airport SAME, Argentina)
SAME APPROACH: ILS RWY36

As a pilot you have to carry out all procedures with the MCDU inserting all the information and flight data to complete the cockpit preparation before the flight. After that, the flight will be automatically done until reaching the STAR chart of destination where you will need to operate the MCDU again to carry out the approach procedures. In this case you have to be ready for an alternate destination. To plan this situation, it's necessary to set one alternative airport in the INIT page and then, set the F-PLN page with the new procedures. Let's check the charts before initializing the MCDU

121

Standard instrumental departure. Sao Paulo Airport.

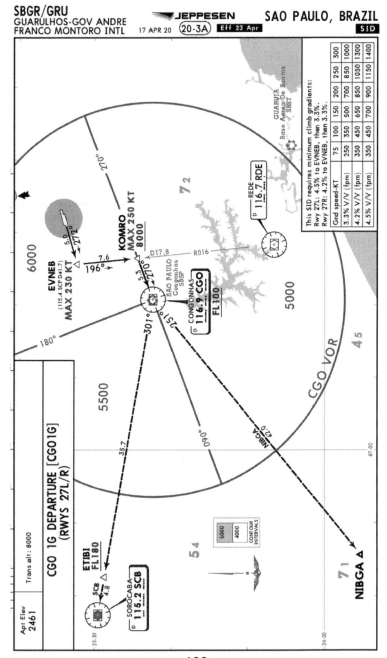

Standard instrumental arrival. Santiago de Chile Airport.

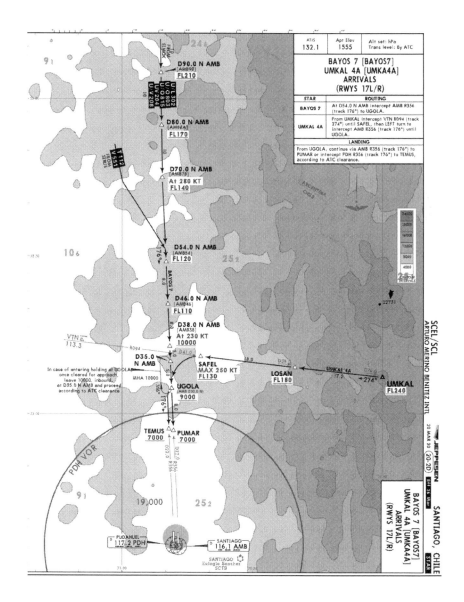

123

ILS Procedure from PUMAR position. Santiago de Chile Airport.

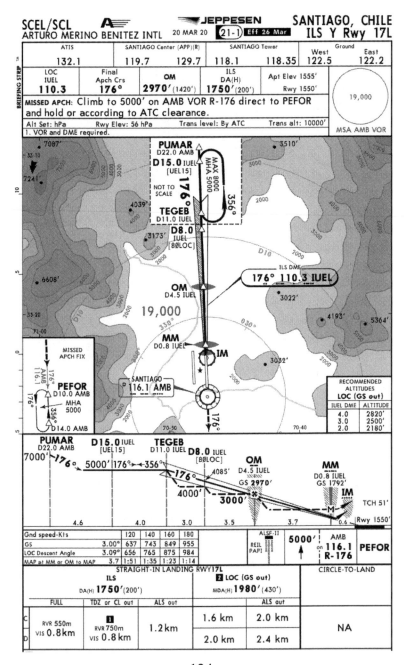

Standard instrumental departure. Santiago de Chile Airport to Mendoza, Argentina (alternate airport)

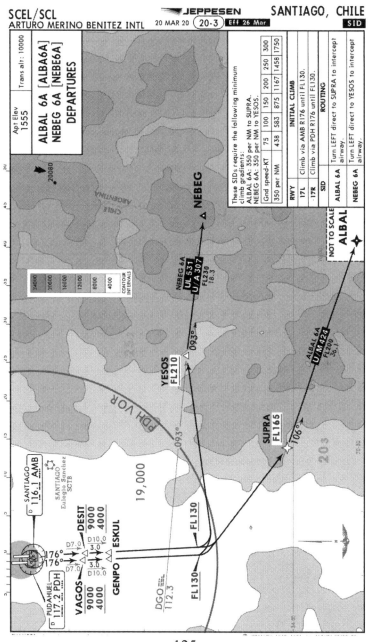

ILS Procedure. Mendoza Airport. (alternative)

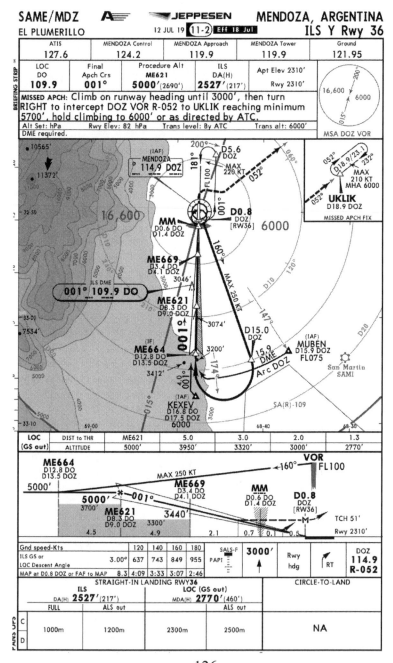

ATIS	MENDOZA Control	MENDOZA Approach	MENDOZA Tower	Ground
127.6	124.2	119.9	119.9	121.95

LOC DO 109.9	Final Apch Crs 001°	Procedure Alt ME621 5000'(2690')	ILS DA(H) 2527'(217')	Apt Elev 2310' Rwy 2310'

MISSED APCH: Climb on runway heading until 3000', then turn RIGHT to intercept DOZ VOR R-052 to UKLIK reaching minimum 5700', hold climbing to 6000' or as directed by ATC.

Alt Set: hPa   Rwy Elev: 82 hPa   Trans level: By ATC   Trans alt: 6000'
DME required.

MSA DOZ VOR

LOC (GS out)	DIST to THR	ME621	5.0	3.0	2.0	1.3
	ALTITUDE	5000'	3950'	3320'	3000'	2770'

Gnd speed-Kts		120	140	160	180		3000'	Rwy hdg	RT	DOZ 114.9 R-052
ILS GS or LOC Descent Angle	3.00°	637	743	849	955	PAPI				
MAP at D0.8 DOZ or FAF to MAP	8.3	4:09	3:33	3:07	2:46					

STRAIGHT-IN LANDING RWY36				CIRCLE-TO-LAND
ILS DA(H) 2527'(217')		LOC (GS out) MDA(H) 2770'(460')		
FULL	ALS out	FULL	ALS out	
C	1000m	1200m	2300m	2500m	NA
D					

126

It's the first flight of the day. The aircraft is in "cold and dark" mode. When the pilot gets in the cockpit, he turns the systems on. The MCDU is off. To turn it on it is necessary to set the BRT knob on, it is located on the right side of the MCDU as you can see in the next picture:

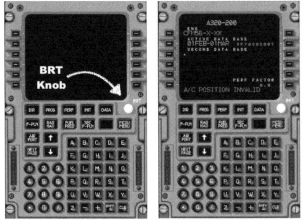

The other systems are on but in standby mode until the pilot inserts the information in the MCDU. The PFD (primary flight display) does not show any information. The ND (navigation display) shows the message MAP NOT AVAIL and the FCU (flight control unit) is ready to set all parameters.

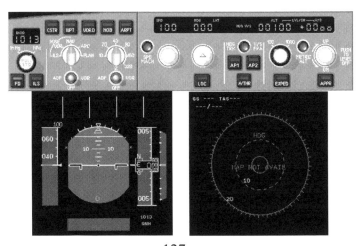

The next step is to initialize the system from the INIT page. Here, the pilot must insert all information about departure and arrival airport, route, flight level, temperature, wind, cost index, and flight number. To do that, the pilot uses the alphanumeric keys to write each data and the lateral keys to insert them in the correct place, as you can see in this picture

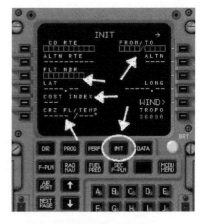

For this example, it's needed to insert SBGR/SCEL (ICAO codes) in the section of FROM/TO. Each word that the pilot writes appear at the bottom of the display but they are not inserted yet and the system remains without information until the pilot inserts it in the correct place.

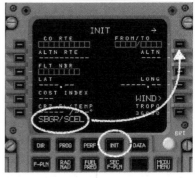

Before aligning the IRS system, the pilot has to establish one alternative airport. In this case the alternative airport will be the international airport of Mendoza City in Argentina (SAME). To load this alternative airport the pilot has to insert it with the lateral key in the section ALTN.

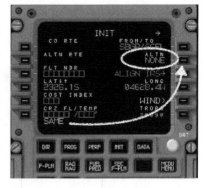

128

After inserting the flight data, the message ALIGN IRS (inertial systems) appears. This is necessary to set the correct position of the aircraft over the airport. When the IRS system is initialized, the PFD and the ND show the initial information. The ND shows the ICAO code and the distance to the destination airport.

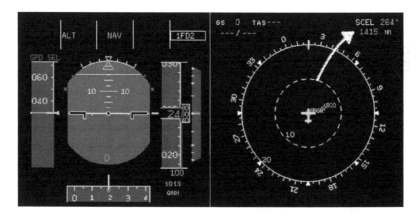

When the INIT page is completed, it shows the information to check and if it's correct, the pilot can continue with the next page. Remember, if there's any error, just write the correct data and insert it again in the correct place. The wrong information will be replaced by the correct information.

The next page is the INIT B. Here, the pilot must set the information about weight and fuel. To access to the INIT B page from the INIT A page, just press the NEXT PAGE key. When this configuration is ready (ZFW +FUEL), the INIT B page shows the value of takeoff weight and landing weight in the right side of the display. In this example TOW 73.3 and LW 65.7.

In this example, the runway in use is 27R and the SID to use is CGO 1. The pilot has to select this option and insert it. Before inserting the SID, its track will be shown with a dashed line on the ND to confirm the correct courses. The system is ready. The next step will be to set the navigation from the F-PLN page. This page shows each point of the route, from departure airport to destination airport and every waypoint in between.

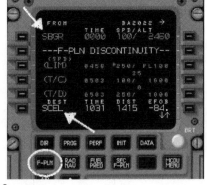

The pilot just needs to set the SID (standard instrument departure), the STAR (standard arrival), the approach chart to the runway in use for the destination and, in this case, the SID and the Approach chart for the alternative plan with Mendoza Airport (SAME). In this example appears the message F-PLN DISCONTINUITY due that the SID is not loaded yet. In order to select the SID for runway in use it is necessary to press the L1 key and then the next page of LATERAL REVISION with all the options will appear.

In this new page appears the title LAT REV from SBGR (ICAO code) to confirm the correct airport. The key L1 on the left side shows the option DEPARTURE to select the runway for takeoff and its SID.

When you press the DEPARTURE key, the all runway option will appear. Beside them, the runway length. And finally, below the runways the information about the ILS.

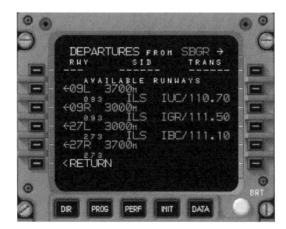

In this example, the runway in use is 27R and the SID to use is CGO 1. The pilot has to select this option and insert it. Before inserting the SID, its track will be shown with a dashed line on the ND to confirm the correct courses.

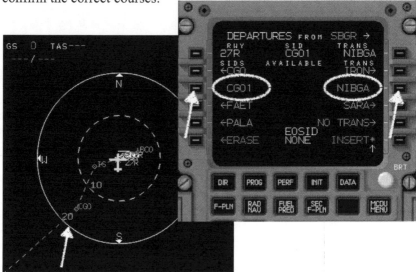

After inserting the SID, its track will be shown with a normal line on the ND, as shown on this picture. Note that the message F-PLAN DISCONTINUITY has disappeared.

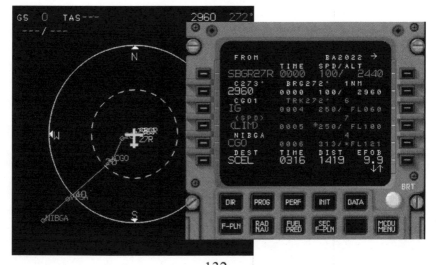

Once everything is loaded, the runway with the correct SID, it's time to select the destination options. To do that you need to press the last key on the bottom of the display where DEST and SCEL (ICAO code of destination) are indicated. After pressing this key, the LAT REV page will appear like on the previous step.

When activating this option, the pilot can select not only the Approach procedure for any runway, but also the STAR chart.

After pressing the key, a new window appears with the option ARRIVAL on the right side indicating LAT REV from SCEL to confirm the correct airport.

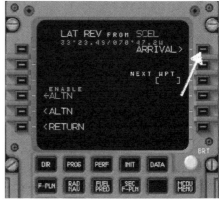

The next page shows the available runways and the IFR procedures for each runway with its data. For this example, the pilot has to select the runway 17 and an ILS procedure. This option is ILS17.

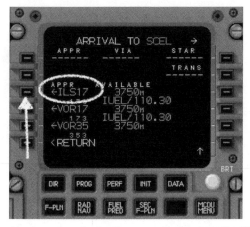

After selecting the desired runway, the STAR CHART options appear. In this case the correct option is UMKAL4. Before inserting the selection, the ND shows the track to fly with a dashed line. If it's correct, the pilot has to confirm and insert this selection.

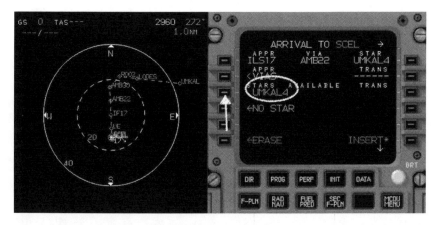

After inserting the STAR CHART and the APPROACH CHART, its track will be shown with a normal line on the ND, as shown on this picture.

Additionally, on the F-PLAN page, the pilot has to set all procedures for the alternative airport, SID from SCEL and Approach for SAME. To do that, it's necessary to scroll down passing the destination with the arrow's keys until reaching the legend END OF PLAN, below it will appear the origin and the new destination.

At this point the flight plan has ended, but the new one begins, so it's necessary to load the new departure and the arrival information. In this new page appears the title LAT REV from SCEL (ICAO code) to confirm the correct airport (like before). The key L1 on the left side shows the option DEPARTURE to select the runway for takeoff and its SID. In this case the same runway for approaching in the original plan.

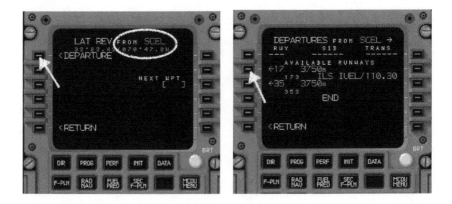

In this example, the SID in use is ALBAL 6A. The pilot has to select this option and insert it. Before inserting the SID, its track will be shown with a dashed line on the ND to confirm the correct courses.

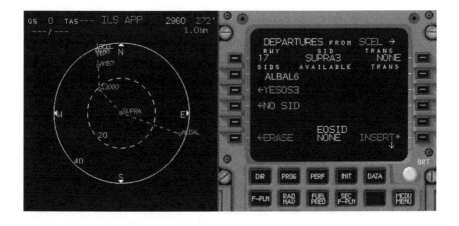

After inserting the SID, its track will be shown with a dashed line on the ND too but in color blue, as shown on this picture.

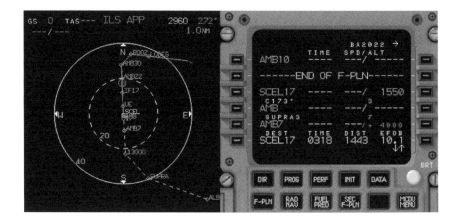

Once everything is loaded, it's time to select the destination options. To do that it's necessary to press the last key where SAME (ICAO code of destination) is indicated and in the same way that the original departure, after pressing this key, the LAT REV page will show.

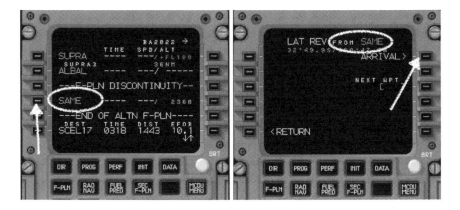

In this example, the pilot will select an ILS procedure for runway 36 of Mendoza airport and the ND will show this procedure in blue.

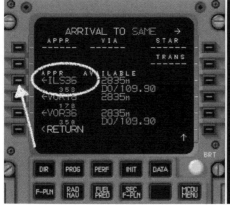
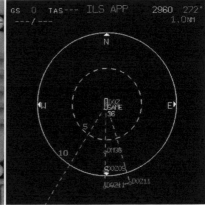

Once the F-PLAN page is completed, the next page will be RAD NAV where the pilot has to set the Nav aids information based on the standard instrument departure chart (SID) to fly. In this case, CGO1 SID. This page gives the option to set two VOR, one ILS and two ADF frequencies, each one with its courses to fly. In this example, the pilot has to select the VOR CGO frequency without any ILS frequency. In the PROG (progress) page, the pilot has to check the optimum and maximum flight level. Additionally, the crew can select any waypoint or any nav aid to obtain the bearing and distance to this reference at any

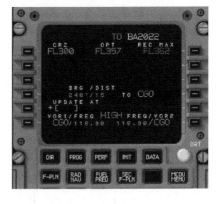

moment during the flight. In this example, the optimum flight level is

138

FL357 and the maximum flight level is FL362. Additionally, the CGO VOR is selected to obtain bearing and distance.

The next is one of the most important pages. In PERF (performance) page, the pilot has to set speeds, altitudes, flaps setting and the takeoff power. In this example, the pilot has set the speeds (V1:125kt. VR: 128kt. V2:140kt), flaps setting at position 3 and the takeoff power with a value of 50°.

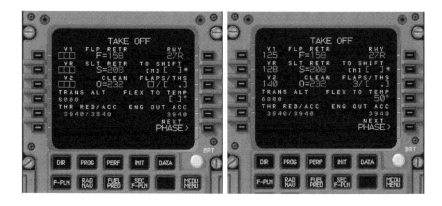

Finally, there is an option that enables the pilot to set a secondary flight plan, but in an inactive mode. This option is useful in case of an emergency where it is necessary to change the flight plan immediately. In this page SEC F-PLN, the system shows three options, copy the active flight plan, create a new one, and set the INIT page for this new flight plan. In this case, the pilot just copies the active flight plan.

When the system is ready and all takeoff data is loaded, the PFD and ND show the information of this phase. The PDF shows information about active speed, altitude and pitch. The ND will show the track line of the SID to fly.

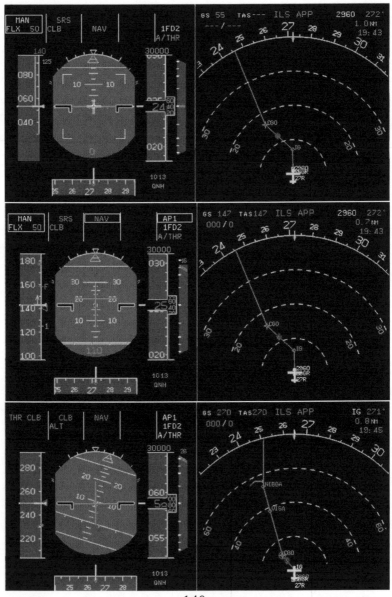

Once the aircraft is flying the SID, on the ND appears a blue arrow in the middle of the route track after NIBGA position. This symbol indicates the TOC or Top of Climb. At this point the aircraft will change from climb phase to cruise phase.

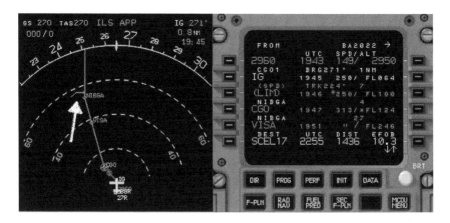

After passing NIBGA position, the route will remain for several miles until the next waypoint. For this example, we will assume that the aircraft is near to destination at 96 NM from UMKAL position, the next waypoint prior to approach. In the MCDU the message ENTER DEST DATA appears indicating that the pilot must insert the destination data.

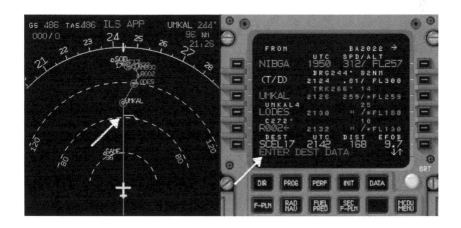

141

Here the pilot must set again the parameters and information in some pages in the MCDU to be ready for initiation of the descent, the approach and landing. The first page is RAD NAV to set the nav aids frequencies. In this example, the frequency of AMB VOR and IUEL ILS.

The next page is PROG where the pilot can set some nav aid or waypoint to get the bearing and distance to this reference. It's not necessary but it's very helpful. In this case, the reference will be AMB VOR.

The aircraft is approaching to the UMKAL position and near to the TOD or Top of Descent. This point is shown with a white arrow in the ND over the route line. In this case, just a few miles before UMKAL. At this point of the navigation, the pilot needs to change the flight phase from cruise to descent and approach phase. To do that, in the PERF page appears the option ACTIVATE APPR PHASE.

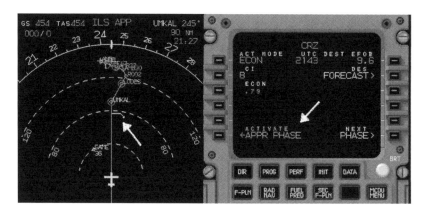

When the pilot activates this option, the aircraft will change the phase and the PERF page will change too. Now in this page are available the destination options like QNH, TEMP, WIND, DH/DA, FLAPS CONFIGURATION and more. In this case, qnh 1013, temp 12°, wind 150°/09, dh 200, flaps conf 3, for ILS17 as planned.

Additionally to the approach information, in the FUEL PRED

page there is some information about
fuel remaining at destination and
information about the extra time with
this fuel. In this case, after landing,
the fuel remaining will be 9.9 tons,
and if the flight has to divert to the
alternative airport, the fuel remaining
after landing will be 8.3 tons.

In the secondary flight plan
page, the pilot just copies the active
plan. In case that the primary flight
plan fails, the secondary flight
plan will be ready with the same approach
procedure. It's not usual that kind of
fail but it's a good technic to always
copy the active plan if there is no
other option.

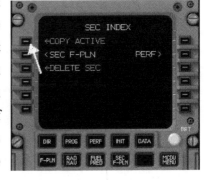

Once the aircraft has initiated the descent, a new restriction
appears over the UMKAL position, it's an altitude's constraint of
FL260. This constraint is shown in the ND with a purple circle and the
F-PLN page represents it with an asterisk between Mach number and
FL.

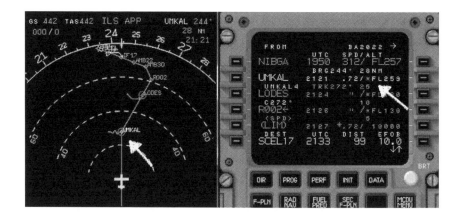

After passing the UMKAL position, the aircraft will continue the descent path, but only to FL180 until LODES position due to another altitude constraint at this point.

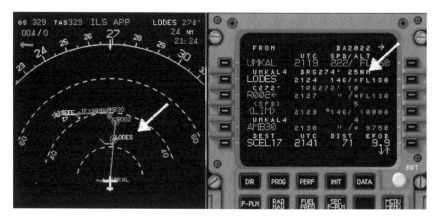

These kinds of constraints could cause that the aircraft flies with a vertical deviation above its descent path. The pilot has to check this vertical deviation from the PROG page.

After passing the LODES position, reaching the final approach, the aircraft continues above its descent path. If the vertical deviation continues increasing, the aircraft will be unable to capture the GS signal, so the pilot has to reduce the vertical deviation as soon as possible.

The recommended technique to reduce the vertical deviation is a smooth increase of speed and the aircraft will increase its rate of descent until reaching the vertical deviation zero or the correct descent path. After increasing the speed, the vertical deviation will reduce. The pilot has to check it in the PROG page. The ND will show two blue arrows, the first one indicates the point where the vertical deviation will be zero and the other will indicate the next descent point.

To carry out the requested by the air traffic controller, the pilot has to activate the secondary flight plan in which the approach procedure is ready for runway 12.

The aircraft is already flying the final approach to the runway 17L at 24 NM from the airport.

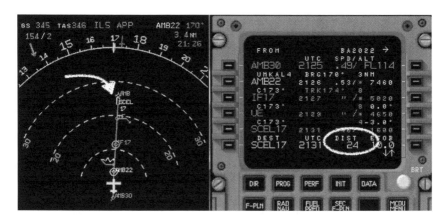

Finally, the aircraft captures the ILS signal and it will continue approaching to the runway. The F-PLN page shows the runway (SCEL17) and after that the next two points in blue for the go around procedure. The ND shows the go around track with a blue line too.

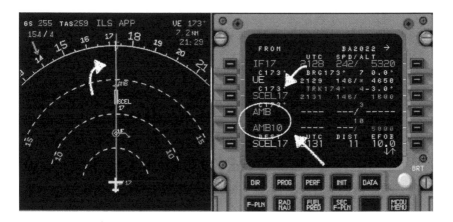

For this example, we will suppose that the weather on SCEL is adverse and the airport has closed just a few NM before landing. In this case the pilot has two options, the first one is to cancel the approach and climb to fly a holding pattern over some point until the airport is open again. And the second option is to divert the flight to the alternate airport, in this case Mendoza international airport (SAME) in Argentina. For this example, the pilot decides to activate the alternate airport. To do that he needs to press the lateral key over SCEL17, then enable the alternate option.

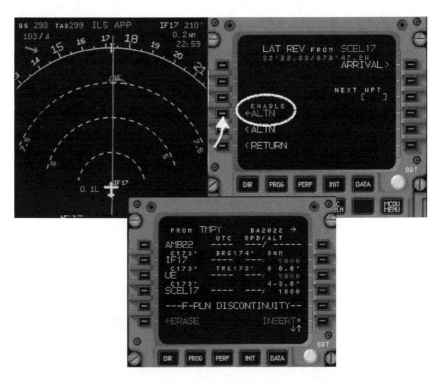

After inserting the new plan, the ND shows the SID track to SAME, the F-PLN page shows new waypoints and the destination has changed from SCEL17 to SAME36.

ALBAL position is the end of the SID. After passing ALBAL, the aircraft will fly directly to DOZ VOR to carry out an ILS procedure.

After several miles, the aircraft initiates the descent. The airport is straight ahead at 90NM. The pilot checks the ILS procedure chart and sets the MCDU with the updated information.

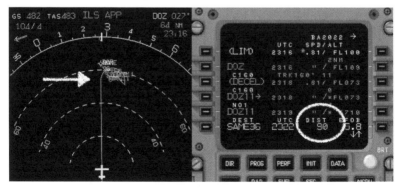

The ILS procedure for SAME runway 36 has a holding pattern over DOZ VOR, then an outbound radial (course 160°) until 15NM and finally the procedure turn to the final approach with the ILS signal. This is what the aircraft has to do.

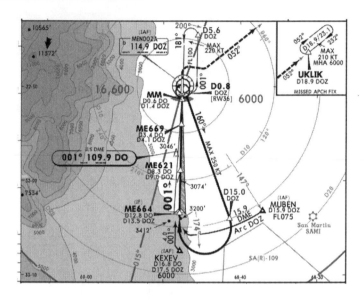

Reaching the airport at about 30 NM, the pilot has to activate the Approach Phase on the PERF page and set the information for approaching.

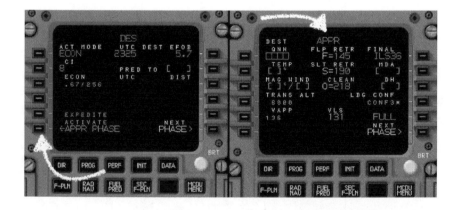

The information for approaching is loaded and the ND shows the diagram of the ILS procedure.

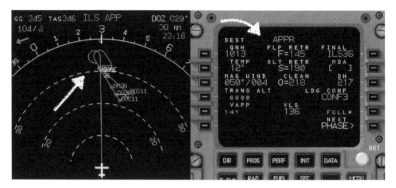

The aircraft will continue approaching to DOZ VOR and the next step will be to carry out the holding pattern procedure. The ND shows this pattern and the F-PLN page indicates HOLD L (left).

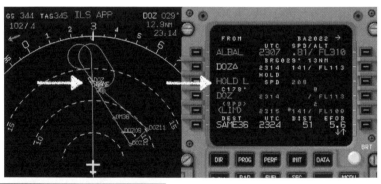

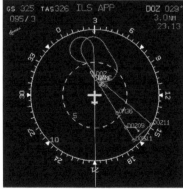

Another option to check the procedure's diagram is the ROSE NAV view on the ND. This view mode lets you check the diagram behind the plane. In the ARC mode it's not possible because this mode only shows diagrams in front the aircraft.

After flying the holding pattern, the pilot has to finish it pressing the option with the lateral key and then he has to continue with the ILS approach until the next point (DOZ 11).

The aircraft continues descending to the next point and after that will turn right until the final approach.

The ROSE NAV mode lets the pilot to keep the airport in sight at all time.

Finally, the aircraft has captured the ILS signal and is flying the final approach prior to land. At this point the flight continues with the autopilot engaged but the pilot can disconnect it at any time to land manually. The ND shows a blue line after the airport indicating the go around track. The F-PLN page indicates this information with letters in blue.

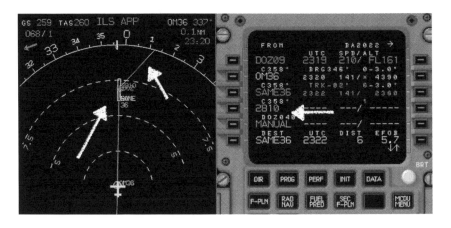

Chapter 5

International Flight. Emergency after take off

(MEXICO to MIAMI)

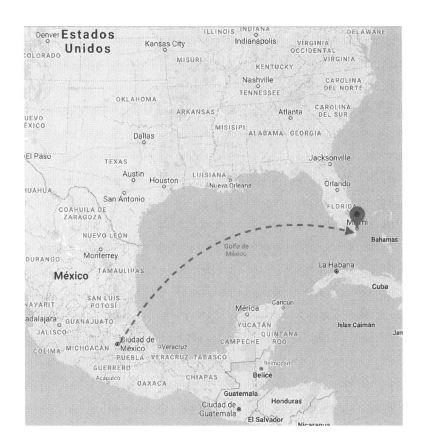

This is for training and entertainment only. For real flight, please see the Airbus manuals.

International Flight II. Emergency after take off

In this chapter we are going to study the work of the MCDU in a real flight with all the phases involved. It's very important that pilots understand each phase and each step of how the MCDU works. If pilots work with the MCDU correctly, the aircraft will flight correctly, but, if the pilots make mistakes working with the MCDU, the aircraft will follow the wrong instructions.

This flight takes place in Mexico from Mexico City (MMMX) to Miami City (KMIA). The aircraft is an Airbus A320. The charts for this flight are the following. In this case, there is an emergency after taking off and the pilot will have to return to the airport as soon as possible, so the Miami's charts will not be considered.

Flight Plan:
MMMX SID: APN 3A
KMIA STAR: no apply
KMIA APPROACH: no apply
MMMX APPROACH: ILS RWY 05R

As a pilot you have to carry out all procedures with the MCDU inserting all the information and flight data to complete the cockpit preparation before the flight. After that, the flight will be automatically done. After taking off until reaching the STAR chart of destination where you will need to operate the MCDU again to carry out the approach procedures. After taking off, the aircraft will have a hydraulic failure and it won't be able to continue to Miami. As a pilot you will have to operate the MCDU to return immediately to the airport. Let's check the charts before initializing the MCDU.

Standard instrumental departure. Mexico Airport.

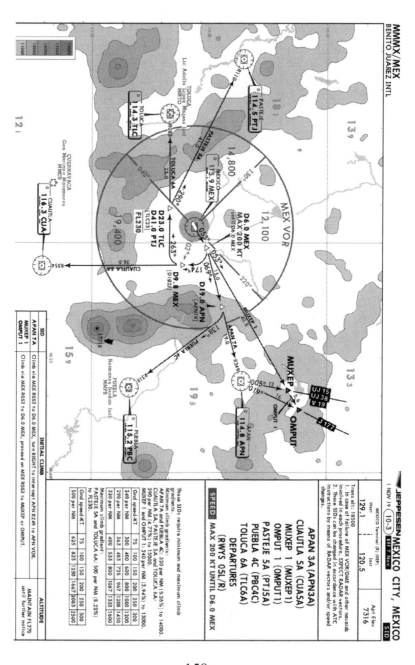

158

ILS Procedure runway 05R. Mexico Airport.

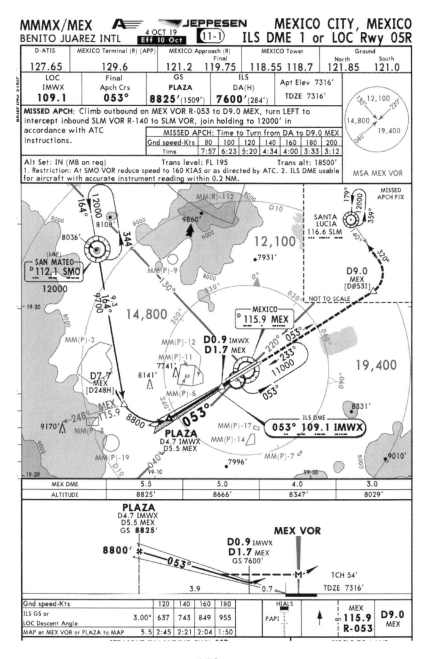

MMMX/MEX — JEPPESEN — MEXICO CITY, MEXICO
BENITO JUAREZ INTL — 4 OCT 19 Eff 10 Oct (11-1) — ILS DME 1 or LOC Rwy 05R

D-ATIS	MEXICO Terminal (R) (APP)	MEXICO Approach (R) Final	MEXICO Tower	Ground North South
127.65	129.6	121.2 119.75	118.55 118.7	121.85 121.0

LOC IMWX 109.1	Final Apch Crs 053°	GS PLAZA 8825'(1509')	ILS DA(H) 7600'(284')	Apt Elev 7316' TDZE 7316'

MISSED APCH: Climb outbound on MEX VOR R-053 to D9.0 MEX, turn LEFT to intercept inbound SLM VOR R-140 to SLM VOR, join holding to 12000' in accordance with ATC instructions.

MISSED APCH: Time to Turn from DA to D9.0 MEX

Gnd speed-Kts	80	100	120	140	160	180	200
Time	7:57	6:23	5:20	4:34	4:00	3:33	3:12

Alt Set: IN (MB on req) — Trans level: FL 195 — Trans alt: 18500'
1. Restriction: At SMO VOR reduce speed to 160 KIAS or as directed by ATC. 2. ILS DME usable for aircraft with accurate instrument reading within 0.2 NM.

MSA MEX VOR

MEX DME	5.5	5.0	4.0	3.0
ALTITUDE	8825'	8666'	8347'	8029'

PLAZA
D4.7 IMWX
D5.5 MEX
GS 8825'

MEX VOR

8800' ✴

D0.9 IMWX
D1.7 MEX
GS 7600'

053°

TCH 54'
TDZE 7316'

3.9   0.7

Gnd speed-Kts		120	140	160	180	HIALS		MEX an 115.9 R-053	D9.0 MEX
ILS GS or LOC Descent Angle	3.00°	637	743	849	955	PAPI	↑		
MAP at MEX VOR or PLAZA to MAP		5.5	2:45	2:21	2:04	1:50			

159

It's the first flight of the day. The aircraft is in "cold and dark" mode. When the pilot gets in the cockpit, he turns the systems on. The MCDU is off. To turn it on it is necessary to set the BRT knob on, located on the right side of the MCDU as you can see in the following picture:

The other systems are on but in standby mode until the pilot inserts the information in the MCDU. The PFD (primary flight display) does not show any information. The ND (navigation display) shows the message MAP NOT AVAIL and the FCU (flight control unit) is ready to set all parameters.

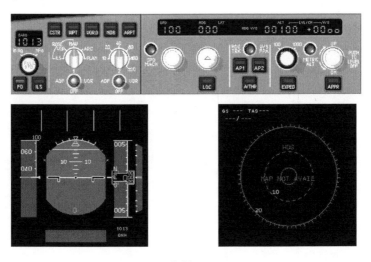

The next step is to initialize the system from the **INIT page**. Here, the pilot must insert all information about departure and arrival airport, route, flight level, temperature, wind, cost index, and flight number. To do that, the pilot uses the alphanumeric keys to write each data and the lateral keys to insert them in the correct place, as you can see in this picture.

For this example, it's necessary to insert MMMX/KMIA (ICAO codes) in the section of FROM/TO. Each word that the pilot writes appears at the bottom of the display but isn't inserted yet and the system continues without information until the pilot inserts it in the correct place.

After inserting the flight data, the message ALIGN IRS (inertial systems) appears. This is necessary to set the correct position of the aircraft over the airport. When the IRS system is initialized, the PFD and the ND show the initial information. The ND shows the ICAO code and the distance to the destination airport.

161

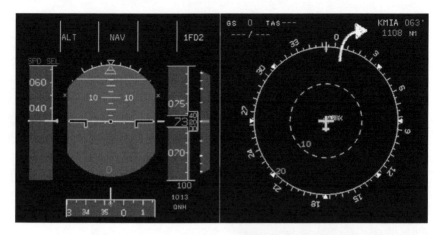

When the INIT page is completed, it shows the information to check and if it's correct, the pilot can continue with the next page. Remember, if there's any error, just write the correct data and insert it again in the correct place. The wrong information will be replaced by the correct information.

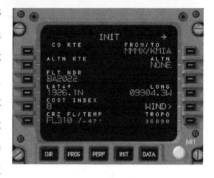

The next page is the INIT B. Here the pilot must set the information about weight and fuel. To access to the INIT B page from the INIT A page, just press the NEXT PAGE key. When this configuration is ready (ZFW+FUEL), the INIT B page shows the value of takeoff weight and landing weight in the right side of the display. In this example TOW 73.3 and LW 65.7.

162

The system is ready. The next step will be to set the navigation from the F-PLN page. This page shows each point of the route, from departure airport to destination airport and every waypoint in between them. The pilot just needs to set the SID (standard instrumental departure),

the STAR (standard arrival) and the approach chart to the runway in use for the destination. In this case it does not apply. In this example appears the message F-PLN DISCONTINUITY due that the SID is not loaded yet. In order to select the SID for runway in use it is necessary to press the L1 key and then, the next page of LATERAL REVISION with all the options will appear.

In this new page appears the title LAT REV from MMMX (ICAO code) to confirm the correct airport. The key L1 on the left side shows the option DEPARTURE to select the runway for takeoff and its SID.

When you press the DEPARTURE key, the all runway's option will appear. Beside them, the runway length. And finally, below the runways the information about the ILS.

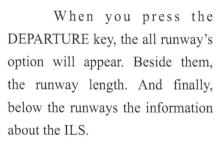

163

In this example, the runway in use is 05R and the SID to use is APAN 3A. The pilot has to select this option and insert it. Before inserting the SID, its track will be shown with a dashed line on the ND to confirm the correct courses.

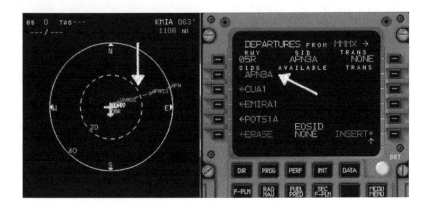

After inserting the SID, its track will be shown with a normal line on the ND, as shown on this picture. Note that the message F-PLAN DISCONTINUITY has disappeared.

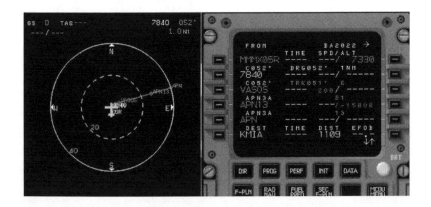

Once everything is loaded, the runway with the correct SID, it's time to select the destination options. Only with training purpose, the pilot loads the destination information. To do that it's necessary to press the last key at the bottom of the display where DEST and KMIA (ICAO code of destination) are indicated. After pressing this key, the LAT REV page will appear as in the previous step.

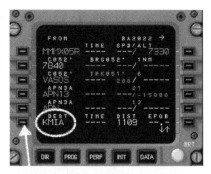

When activating this option, the pilot can select not only the Approach procedure for any runway, but also the STAR chart. After pressing the key, a new window appears with the option ARRIVAL on the right side indicating LAT REV from KMIA to confirm the correct airport.

The next page shows the available runways and the IFR procedures for each one with its data. For this example, the pilot has to select the runway 09L and an ILS procedure. This option is ILS09L.

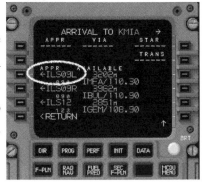

After selecting the desired runway, the STAR CHART options appear. In this case the correct option is DVALL 1 trans EYW. Before inserting the selection, the ND shows the track to fly with a dashed line. If it's correct, the pilot has to confirm and insert this selection.

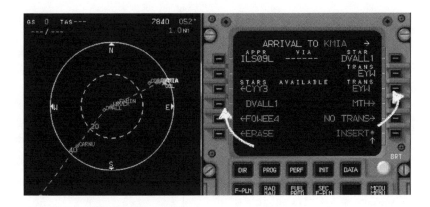

After inserting the STAR CHART and the APPROACH CHART, its track will be shown with a normal line on the ND, as shown in this picture.

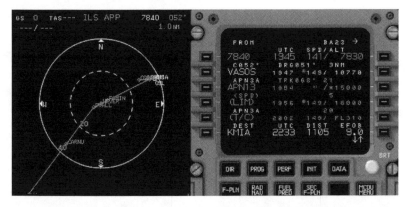

Once the F-PLAN page is completed, the next page will be RAD NAV where the pilot has to set the Nav aids information based on the standard instrument departure chart (SID) to fly. In this case, APN3A SID. This page gives the option to set two VOR,

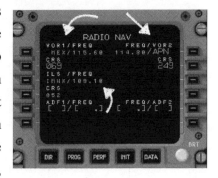

an ILS and two ADF frequencies, each one with its courses to fly. In this example, the pilot has to select the VOR MEX and APN frequency and the ILS frequency.

In the PROG (progress) page, the pilot has to check the optimum and maximum flight level. Additionally, the crew can select any waypoint or any nav aid to obtain the bearing and distance to this reference at any moment during the flight. In this example, the optimum flight level is FL362 and the maximum flight level is FL367. Additionally, the MEX VOR is selected to obtain bearing and distance.

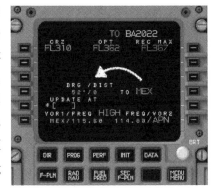

The next is one of the most important pages. In PERF (performance) page, the pilot has to set speeds, altitudes, flaps setting and the takeoff power. In this example, the pilot has set the speeds (V1:123kt. VR: 130kt. V2:135kt), flaps setting at position 3 and the takeoff power with a value of 50°.

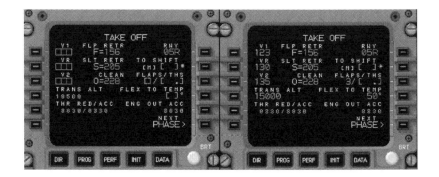

Finally, there is an option that enables the pilot to set a secondary flight plan but in an inactive mode. In this page SEC F-PLN, the system shows three options: copy the active flight plan, create a new one and set the INIT page for this new flight plan. In this case, the pilot just copies the active flight plan.

When the system is ready and all takeoff data is loaded, the PFD and ND show the information of this phase. The PFD shows information about active speed, altitude and pitch. The ND will show the track line of the SID to fly.

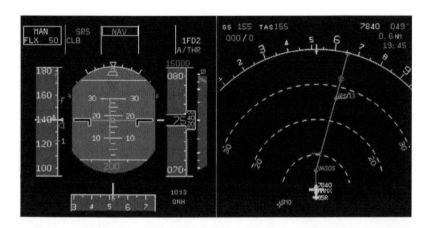

Everything is normal. After taking off, the aircraft flies the SID and the pilots are checking the instrument panel to confirm the correct operation.

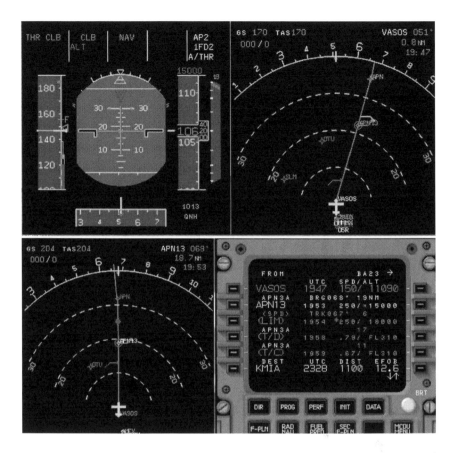

The flight continues climbing to FL310, everything is normal and suddenly the MASTER WARNING alarm is triggered! Hydraulic Failure! The pilot declares the emergency with the air traffic control and requests radar vectors for an immediate return.

Pilot: *Mex control, MAY DAY, MAY DAY, MAY DAY! We have a hydraulic failure, request radar vector to holding for technical checks!*

Control: *Roger, turn left direct to SLM VOR. Approach in use ILS runway 05R from SMO VOR.*

169

At this point, the flight will not continue to Miami and the pilot has to change the destination airport on the MCDU. To do so, there are two options. The first one is pressing the lateral key of destination, then the ALTN option to enable an alternate airport.

After inserting the new destination as an alternative airport, the option ENABLE ALTN appears on the LAT REV page. The pilot just has to enable the alternate and the aircraft will change the flight plan. After that, the new arrival information will be required, ILS information, weather information and landing   performance information.

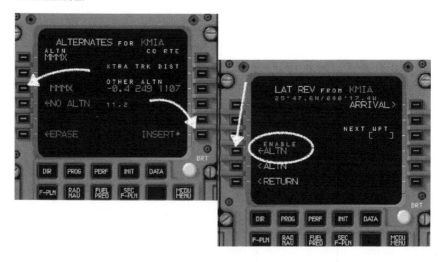

The second option is faster than the first option. To change the destination airport, the pilot has to press any lateral key and insert the new airport in the option NEW DEST.

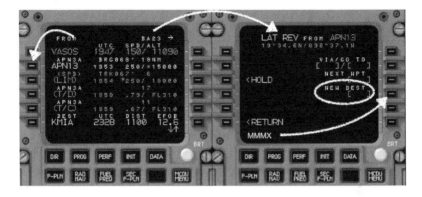

The pilot takes the second option and sets MMMX as the new destination, then the aircraft turns left to SLM VOR. On the F-PLN page the message F-PLN DISCONTINUITY appears because there is not an arrival procedure loaded yet.

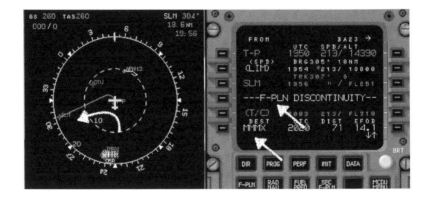

Once the aircraft is heading to the SLM VOR, the pilot sets the approach procedure for runway in usc. In this case ILS05R.

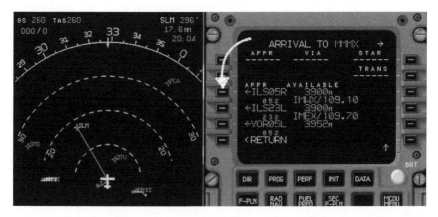

The ILS procedure is via SLM VOR. Before inserting the procedure, the pilot has to confirm it with the ND diagram (dashed line).

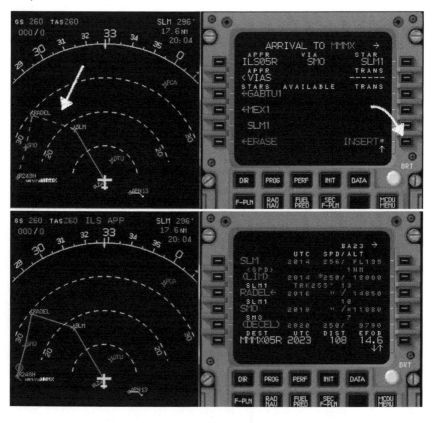

At this point it is necessary to carry out some procedures about the failure. To do so, the pilot requests a holding pattern over SLM VOR. To insert a holding pattern, the pilot has to press the lateral key of SLM and then press the hold option.

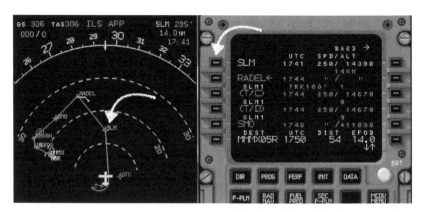

Before inserting the holding pattern, the ND shows a diagram with a dashed line to confirm the correct pattern.

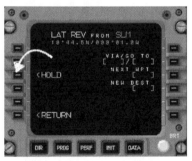

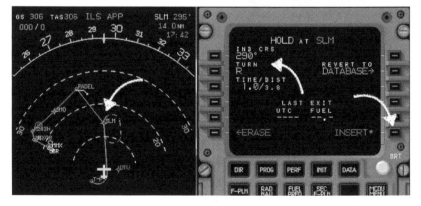

Once the aircraft is in the holding pattern, the pilot has to activate the approach phase to reduce speed and load the approach information. Some technical checks are carrying out while the aircraft is holding over SLM VOR.

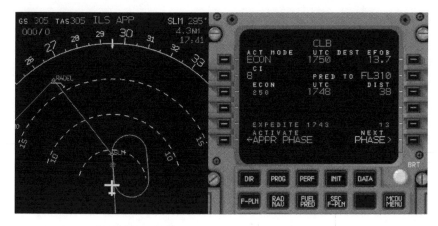

Prior to continue the approach, in the PERF page the pilot has to set the approach information about wind, DH and flaps setting.

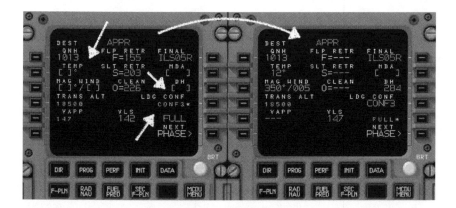

The aircraft is flying the holding pattern. All checks about the failure has been done. Now it's time to continue with the approach procedure to the runway in use. The pilot calls to the air traffic controller and requests to fly directly to SMO VOR.

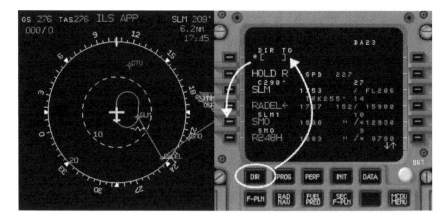

After pressing the DIR key, the option DIR TO appears. Here the pilot has to select SMO with the lateral key and the aircraft will fly directly to SMO VOR.

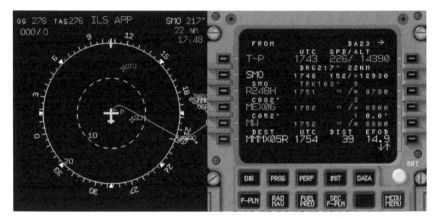

According to the approach chart, there is another holding patter over SMO VOR, so the pilot has to set it pressing the lateral key over SMO.

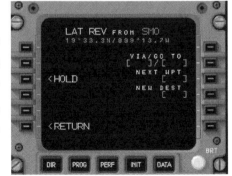

175

The emergency situation is getting worse and there is no time to fly another holding pattern, sow the pilot decides to disregard this procedure and continue with the approach. In this case it's necessary to clear the holding patter from the flight plan. To do that, the pilot just has to press the CLR key and then press the lateral key of HOLD L.

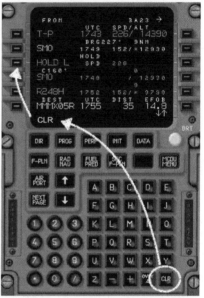

After passing the SMO VOR, the aircraft initiates the final descent in accordance with the approach chart to the next point prior to capture the ILS signal.

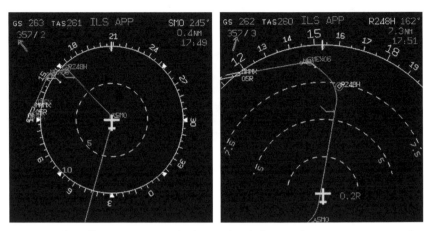

The pilot requests priority to land due to the emergency, so the other aircrafts are holding near the airport to give way. The aircraft is reaching the final turn.

Finally, the aircraft is flying the ILS procedure in the final approach.

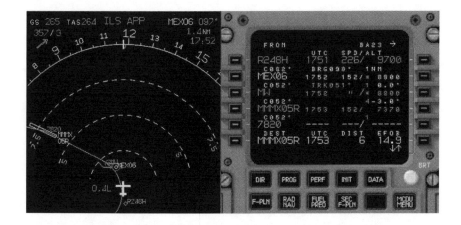

After the last turn, the flight continues with ILS references. The runway 05R is straight ahead and the pilot decides to disconnect the autopilot and continue to land manually.

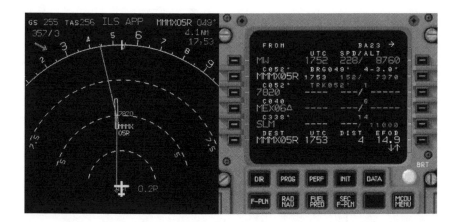

Made in the USA
Columbia, SC
18 August 2024

40682584R00109